THE OTHER DARK MATTER

THE OTHER DARK MATTER

THE SCIENCE AND BUSINESS OF TURNING
WASTE INTO WEALTH AND HEALTH

LINA ZELDOVICH

The University of Chicago Press CHICAGO AND LONDON

The University of Chicago Press, Chicago 60637
The University of Chicago Press, Ltd., London
© 2021 by Lina Zeldovich
Published 2021
Printed in the United States of America

30 29 28 27 26 25 24 23 22 21   1 2 3 4 5

ISBN-13: 978-0-226-61557-8 (cloth)
ISBN-13: 978-0-226-81422-3 (e-book)
DOI: https://doi.org/10.7208/chicago/9780226814223.001.0001

Library of Congress Cataloging-in-Publication Data
Names: Zeldovich, Lina, author.
Title: The other dark matter : the science and business of turning waste
into wealth and health / Lina Zeldovich.
Other titles: Science and business of turning waste into wealth and health
Description: Chicago ; London : The University of Chicago Press, 2021. |
Includes bibliographical references and index.
Identifiers: LCCN 2021013488 | ISBN 9780226615578 (cloth) |
ISBN 9780226814223 (ebook)
Subjects: LCSH: Sewage disposal. | Sewage—Recycling. | Sewage
sludge—Recycling. | Sewage disposal—History. | Sewage sludge as
fertilizer. | Sewage sludge fuel. | Biogas. | Feces—Therapeutic use.
Classification: LCC TD741 .Z45 2021 | DDC 628.3/8—dc23
LC record available at https://lccn.loc.gov/2021013488

♾ This paper meets the requirements of
ANSI/NISO Z39.48-1992 (Permanence of Paper).

To my grandfather
who taught me to plant, grow, and harvest
To my parents
who forgot to teach me how to give up
To Dennis
who believed in me more than I believed in myself
To Tanya
who told me I could write a book
To Luba
who took on my unconventional idea
To my children
who put up with me while I wrote it

CONTENTS

PART 3: THE FUTURE OF MEDICINE AND OTHER THINGS

PART 1
THE HISTORY OF HUMAN WASTE

CHAPTER 1

HOW I LEARNED TO LOVE THE EXCREMENT

Every fall, my grandfather would set off to do two things: prepare our small family farm for the long Russian winter and empty out our septic system.

He would don his sturdy overalls and heavy-duty gloves, both gray as the rainy sky above, tie two old beat-up buckets to long, thick ropes, and head out to the sewer tank buried by the fence surrounding our property. As he opened the pit, ringed with garden weeds and stinging nettles, the stink slowly wafted through the air, settling over our land like a small stomach-turning cloud. Yet my nose would get used to it, and soon I wouldn't be bothered by it at all. I even wanted to help, but my grandmother wouldn't let me go near the pit—she was afraid I'd fall in.

Today's septic tanks can last for years without being emptied, but ours would fill up much more quickly. We could have called a service to empty it, but my grandfather wouldn't let all those riches go to waste. He had a system.

Dressed in his hazard suit, he would spend a day or two dipping buckets into the foul brown liquid and distributing it over our land. Sometimes he carried the buckets by hand, sometimes he balanced them on a *koromyslo*—an arched wooden pole placed over the shoulders to distribute weight evenly. He walked slowly to avoid splashing the stuff onto his clothes and boots.

He didn't just dump the sludge wherever. He poked small holes in the strawberry and tomato patches, where the plants were already shriveled, expecting winter—and poured the goo into them, deep into the earth. He dug little trenches around the apple trees and emptied his buckets onto their roots. And he also dumped a bunch into one of the compost pits, adding it to the already accumulated dead plants, leaves, and our kitchen food scraps. The compost pits operated on a rotating schedule. At the end of the season, he'd close up the current one for a couple of years, leaving it to ferment and biodegrade. When he opened it again two seasons later, after the snow melted and planting time arrived, all the original biomass was gone. The pit was full of soft, black dirt teeming with fat, lazy earthworms that crawled around slowly, too heavy from gobbling down all that food. The sewage odor was gone, too. Instead, the pit smelled of rich, fertile soil, nature, spring, and the promise of the next harvest. And it made me hungry because I thought of all the food we were going to grow with it.

And so when the next fall rolled around, the stinky sewage pit didn't disgust me. It was simply a part of life, of the natural cycle necessary to grow food and put dinner on the table. I loved growing everything—strawberries, apples, tomatoes. As I puttered along with Grandpa, he taught me what plants needed to thrive. I liked to sow and I liked to harvest. I even liked to weed. After weeding, the patches looked so clean and orderly that you could just see the fruit starting to burgeon faster. Our farm wasn't big, but it yielded enough veggies and berries to pickle and preserve for the long Russian winters, when nutrients were scarce. And we grew so many apples that they lasted in our cellar through spring.

One of my grandmothers was a doctor, the other a chemist, and my mother and father were working on their dissertations in mechanics and physics, respectively. Before I started school, I knew about the perilous microbes that lurk in the soil, our dog's poop, and our own excrement. That's why I had to wash my hands twice after planting, get the muck out from under my nails, and carefully scrape the garden dirt off my boots before entering the house. Washing everything was essential. We had no hot water in the house, so Grandma would heat up some for me. She had a process for cleaning Grandpa after his sewage excavations, too. He took off his hazardous overalls outside, and Grandma stuck them in a metal vat full of water, which she would later boil on the stove for a half hour until all the pathogens were dead. She didn't cook his boots because that would destroy them, so she just scalded them with boiling water. The germs were taken very seriously and given no chance to spread. When we got sick, it was mostly from the bitter winter cold. Diarrhea was rare. We just didn't get that many stomach bugs. We were quite aware of the dangers of raw sewage.

My grandfather, too, was well educated for his time. He had two college degrees, one in agriculture and one in engineering, so he combined knowledge from both with the wisdom of the generations of farmers in his family. Today, we may call it organic farming, composting, recycling, or the circular economy. But for him it was, simply, the way of life. He did use some inorganic fertilizer from heavy-duty plastic bags that he kept in the shed, but he did so sparingly. Inorganic fertilizers gave plants too strong a boost, and that wasn't a good thing, he thought. Manure, whether cow or human, was a natural way to feed your fields. You fed your land like you fed people.

The Russian equivalent of the word "fertilizer" is *udobrenie*, which comes from the word *dobró*, meaning "good" and "rich." *Udobrenie* meant returning all that good stuff to the earth. Even the common Russian colloquialisms that revolve around the toilet recognize that. When toddlers were being potty trained, we jokingly referred to the moment they had to go as giving out *dobró* or *bogatstvo*—"the riches." I think that's why I never viewed sewage as waste. I always considered it more of a treasure than a nuisance. If we didn't return it to the earth, the earth wouldn't have the stamina to feed us.

I knew that other people, who lived in big apartment buildings, didn't have septic tanks, but as a child I assumed that their sewage was also somehow returned to the earth. If not, what would they eat? What would we all eat if we didn't replenish our earth? We had to feed it just like we fed ourselves. This natural cycle had to keep going forever.

I soon learned otherwise.

The land we lived on wasn't entirely ours. My grandfather received the parcel from the Russian airplane factory where he worked as

FIGURE 1. My grandfather, Lev Abramovich Gurevich, on our small family farm in Kazan, Russia. Every fall, he used our septic system to feed the land that fed us. He spent 30 years cultivating our farm's soil, making it a soft, rich, and moist black gold. CREDIT: SOLOMON ZELDOVICH AND LEO CIERI

an engineer during World War II. In a country ravaged by war and plagued by food shortages, factories wanted their workers to be self-sustaining by growing their own produce. At the time, these plots were on the city's outskirts, more of a village than the city itself.

Thirty years after the airplane factory gave the parcel to my grandfather, the city had grown—and so had the demand for land. The factory took the land back to build apartment buildings to house more workers, who would build more jets. Our village fought back—and lost. The factory moved the displaced families into apartment buildings, demolished their houses, and bulldozed their plots. When the bulldozers roared in, squashing our vegetable patches under their massive steel tracks on a gray autumn day, the sky dropped little beads of water onto us, as if mourning our farm. I wished I could cry with it, but I had no tears left. I just watched our apple trees falling to those angry metal machines like tin soldiers. When they finished, the bulldozers scraped off the rich black topsoil that my grandfather had spent 30 years cultivating and trucked it away to the summer homes of the ruling elite. They wanted not just our land, but our dirt, too. They knew that black dirt was gold. It *was* black gold.

Our new apartments had flushing toilets and hot water, but did little to help us heal. We grieved for our farm. A month after we moved, I was struck with a rare disorder that gradually took my ability to stand straight and walk. I lay in bed, homeschooled by my parents, slowly losing feelings in my legs. My grandfather's legs failed him, too, albeit in a different way—he developed bone cancer. We both collapsed like our apple trees, as if someone had pulled our roots from the rich soil that fed us.

Three years later, experimental surgery put me back on my feet. My grandfather was less fortunate. He died shortly after I came home from the hospital, as if he had waited to see me standing strong like a tree that took well. I never lived on a farm again, but every spring I missed the musky, earthy smell of the open compost pits and the promise of the next harvest.

A few years later, when my family moved to the United States and settled in New York City, I learned quickly that most people did not know where their sewage went. And they had no desire to know, either. They were quite happy to flush and forget. For a long time, Western society believed this was the way to go. Governments and private companies built big sewage plants with the goal of rendering the matter harmless, but not necessarily reusing it. We destroyed the organic goodness we produced, and we forged synthetic fertilizers to grow our crops. And at the same time, the developing countries kept struggling with disease outbreaks and other sanitation issues stemming from fecal contamination of drinking water. It seemed that neither world could get its shit right. Mother Nature's link was clearly broken.

And then, in the past decade, I saw the sewage tide turning back. The cultural taboos shrouding toilet habits, human waste, and sewage disposal began to thin. People became interested in what happened after they flushed. The discussions about recycling various kinds of waste pivoted to excrement. We began to talk about recycling poop. Sanitation engineers began to describe wastewater treatment processes as resource recovery activities. Epidemiologists worldwide used sewage to track disease outbreaks.

In 2011, the Bill and Melinda Gates Foundation issued its Reinvent the Toilet Challenge, asking the world's brightest minds to redesign our privies to recover valuable resources from our waste, including energy, nutrients, and clean water, and to operate autonomously without connections to sewage treatment plants. Scientists and engineers talked about the value of the circular economy that favored the concept of giving back to the earth. Brave and brilliant startups sprouted all over the world, devising new ways to turn waste into wealth. Environmentalists, entrepreneurs, and organic farmers gradually began to see their excrement as *dobró* and *bogatstvo*. They began to recognize that every time they had to go, they gave up riches, which shouldn't be flushed down the drain. And what was still going down the drains proved useful, too. As I write this during the COVID-19 pandemic, scientists worldwide are using sewage samples to discover whether the coronavirus is present in their locales, when it first arrived, and exactly which strain is circulating now. Public health scientists at my own alma mater, Columbia University, have been sampling wastewater at different dormitories to detect traces of the coronavirus's genetic material. Infected people shed the virus in their stool even if they don't have symptoms. Thus, the scientists can alert the university about outbreaks brewing so it can stop the virus from spreading before it takes hold, rather than having to shut down the entire campus.

Poop itself became classy and stylish. In 2017, one of Amazon's best-selling toys was a soft, squeaky, plush poop emoji. The next year, that image became so trendy it made a fashion splash, appearing on shirts, backpacks, hats, and pillows. Sometimes it was rainbow-colored or embroidered with shimmering sequins. That popularity proved remarkably persistent. In 2020, for example, pooping toys and clay-like poop substances were named top Christmas gift ideas for kids.[1] My local pharmacy tempted me with a pair of puffy poop slippers and a matching candy dispenser as holiday treats. A café in Seoul serves its lattes with a creamy white poop emoji on top. And did you know that America now has the National Poop Day? According to EventGuide.com, it takes place on the first day after Super Bowl Sunday, as if publicly celebrating the relief from all the junk food we have eaten the night before. Poop graduated from being gross to being a part of nature and a part of us.

It was as if the world embraced the wisdom that my grandfather followed 70 years ago, when he first built his farm. And that was very exciting. Because once we restore the broken link, we will use less fossil fuel, waste less energy, generate less acid rain, and produce fewer greenhouse gas emissions, fewer bleached corals, and fewer dead fish. Instead, we will clear algal blooms from our waters, slow down global warming, and grow healthier food with the fully organic goodness that we ourselves have produced.

It's one thing we are all good at.

Naturals, we just need to recognize that power within us. Today, in the era of soil depletion, weather extremes, and the planet's growing population, recognizing our own power is especially important. It may not be pretty to look at, but it's a treasure.

CHAPTER 2

THE EARLY HISTORY OF HUMAN EXCRETA

Our nomadic ancestors had it easy. They answered their calls of nature whenever and wherever they felt like it—exactly as any other nonhuman animal on earth does. The early humans had no privacy issues and no preferences regarding toilet paper. They simply relieved themselves where they wanted and wandered away from their droppings, leaving them for Mother Nature to process. And she did.

In fact, back then, Mother Nature dealt with human excrement well. She got rid of the crud in the natural, ecologically sustainable, healthy-for-the-planet way. We may look at our waste as yucky, pathogen-laden filth, but to the great outdoors it's just another organic substance that perfectly illustrates the saying that one man's trash is another man's treasure. That ugly, foul-smelling, disease-spreading substance is full of nutrients, particularly nitrogen, phosphorus, and potassium—the important building blocks for plants and all living things.

Our ancestors had their basic necessities taken care of in the ideal fashion. They were in some ways lucky because they didn't have to deal with processing their waste. They simply walked away from their dangerous deposits and moved along, chasing herds of gazelles and looking for berry bushes. And once their bodies extracted the nutrients from all that food, more deposits would fall onto the earth, keeping the cycle going.

But nothing ideal lasts forever. As the early hominids fanned out from Africa, strolled across Siberia, and walked over the mountains, they fertilized places along the way—especially those "rest stops" where their clans chose to stay for a while. Some of these early humans may have even noticed that plants tended to grow bigger, better, and tastier in such rest stops than they did in other places. So tribes made a point of coming back to those spots the next season, or even several years in a row, to enjoy the good food and fertilize the land again and again. And then, one year or one day, some of them may have decided that this nice fertile spot was just too good to abandon. Besides, that constant walking was tiring, winters were cold, and you never knew what you would have for dinner and where you would spend the night. Staying put, building a shack, saving some of the gathered grain for winter to chew on or cook with—and perhaps helping those tasty plants grow—made a lot more sense.

Those prescient thinkers brought us to the advent of farming. They were the ones who began cultivating land and domesticating animals, who switched from a nomadic lifestyle to a settled way of life. They are the ones we credit with the establishment of modern civilization. They are also the ones we must credit—or blame, depending on the point of view—for leaving humankind forever stuck dealing with its waste. Once idyllic wandering—if sleeping on the ground without pillows, blankets, heating, or air conditioning can be called idyllic today—yielded to homesteading, humans could no longer walk away from their shit. And sure enough, shit began to pile up.

At first, that piling up was probably rather slow. The tribes weren't very large, and people still moved around a fair amount. They walked in search of edible plants and chased deer in the woods, and thus dispersed their excrement reasonably evenly. But as they relied more and more on growing food and keeping farm animals, the problem began to raise its stinky head.

If there's one thing we have in common with our ancestors, it's that they were just as dismayed with their excrement as we are with ours. They didn't have the deep biological and bacteriological knowledge that we do, but they were shrewd enough to realize that having feces lying around wasn't good.

Left outside, a pile of poo becomes a nutritious and tasty dinner to certain species. Drawn to the nitrogen, phosphorus, undigested proteins and carbohydrates, and other energy sources stored in it, insects, helminths, bacteria, and other pathogens move in. All the disgusting flies and maggots you may have observed in the decomposing dung of dogs, horses, or other animals are either feasting on or depositing their eggs in this nutrient-rich medium. Earthworms munch on the dung, converting it into compost, while various bacteria finish off what's left. And so the pile of poo quickly turns into a dangerous health hazard.

Not possessing microscopes, or knowledge about germs and how they spread disease, our ancestors went by their gut. Whether it was because of the smell, the look, or the bugs that swarmed over it, biting and annoying them, even our Neolithic ancestors wanted nothing to do with their shit.

Some dug pits away from their dwellings or in the middle of their fields. Some designated "bathroom spaces" outside the village, or behind the bushes, or underneath the trees. Some went out to the riverbanks, where the excrement would get carried away by water— possibly to the dismay of the villagers living downstream. For as long as the settlements were small, those methods worked. The key was to have that stuff out of your dwelling, as it only exacerbated the already profound lack of comfort. But the little villages kept growing, and some eventually became cities. As populations grew, fields and forests shrank, and space became scarce, all that shit really began to stink. Even the early civilizations found their ordure so bothersome that they got creative about making it disappear. In Skara Brae, a Neolithic village in today's Scotland, archaeologists found "a 5,000-year-old, stone-built drainage channel which connected the house to an outfall at the sea edge. The drains were made of stone and had originally been lined with tree bark to make them water-tight. It was indeed a remarkably sophisticated system for its time."[1] And as humanity grew, the sewage systems began to grow, too.

The Bronze Age Plumbers

During their peak, some of the largest and most successful Bronze Age cities counted thousands of people. The Minoan civilization, which flourished on Crete and other Aegean islands from about 2600 to 1100 BC, pre-dating ancient Greece, had over 100 cities. Knossos, the Minoans' largest city, counted 80,000–100,000 inhabitants in its heyday.[2] An average adult produces about a pound of poo a day, and an average child a little less, so assuming that their healthy, high-fiber Mediterranean diet kept them regular, the Minoans generated nearly 50,000 tons of feces daily, all deposited within a relatively limited physical space and accumulating week after week. Part of it probably ended up fertilizing some vegetable patches nearby, but 50,000 tons a day is more than city gardens can handle. What does one do with all that shit? And what does one use as an easy and readily available force to purge it from yards and homes?

The engineering answer was water. Every human civilization has been located next to some water source—a lake, a sea, or a glacier-fed river—because without water, life, food, and daily human activities don't really work. And so a few smart Minoans, frustrated by the daily battle with their excrement, turned to water as a solution. And they were the ones to credit—or blame, again, depending on the point of view—for teaching humankind to dump its sewage into the water. They were the first to set the precedent of discarding our unwanted excrement into aquatic basins. They were the ones who started flushing waste out into the deep blue, rather than keeping it on land.

This new step in excrement history was very important, not only because it led to the creation of sewage systems as we know them, but also because it began to alter the existing nutritional balance of land and water ecosystems, which left us grappling with many of the environmental problems we are experiencing today.

The health of soil ecosystems has always depended on their having sufficient concentrations of nitrogen, potassium, phosphorus, and carbon, as well as some other nutrients like iron, magnesium, and sulfur. Without these elements, the plants can't build their cell walls or convert carbon dioxide into oxygen. Take these nutrients out of the land, whether it's a forest, a meadow, or an agricultural field, and plants simply won't grow. Add more of these nutrients to your garden patches, whether in the form of compost or some other type of fertilizer, and crops will flourish. The richest soils, coveted by farmers and gardeners worldwide, have always been high in these basic nutrients.

By contrast, aquatic and marine environments have historically been low in these elements. And that's fine. Marine species evolved to require low concentrations of nitrogen, potassium, phosphorus, and other nutrients to stay healthy. For aquatic ecosystems, an over-abundance of these elements isn't a good thing. When fertilizer run-off over-enriches rivers and oceans today, it fuels algal blooms and smothers corals. All plants and animals need nourishment to function, but just the right amount of it. Give us too little, and we starve to death. Give us too much, and our bodies grow out of control, becoming obese.

For as long as nitrogen and phosphorus were dutifully returned to the soil by land creatures, the earth grew plenty of food. And for as long as all that excrement remained on land, rather than being flushed into the water, the lakes and seas continued to be clean and healthy ecosystems, supporting a diverse mélange of inhabitants. The land remained nutrient-rich, the water remained low in nutrients, and the planet remained in an overall balance—albeit a constantly shifting and changing balance, moving through the seasons and recovering from volcanoes, asteroids, and other calamities. Using water to rid us of our excrement, however, began to crucially alter that vital balance, which would later become more and more damaged throughout the progress of civilizations.

The Minoans, who were so concerned about keeping their dwellings and streets clean, figured out ways to collect and direct water and began using it for this purpose. Thus, the Minoan civilization had the first ever, simple yet functioning, version of a flushing toilet and a sewage system.

Four thousand years ago, the Palace of Minos in Knossos had a cleaning system in which rainwater from the roof was gathered and used to flush the sewage from three bathrooms in the east wing. A sophisticated water system directed different sources of wastewater into pipes underneath the floors, which then joined together to form a large underground channel that also disposed of toilet contents.

Every so often, seasonal downpours would flush the sewers clean, but toilet users did so, too. Archaeologists found evidence that palace residents used large water pitchers to flush their toilets. They also preferred sitting on them, not squatting—judging by a stone seat that has survived to our days in perfect condition. Just like our modern plumbing, the Minoan pipes occasionally clogged, so the underground sewers came equipped with manholes for cleaning, maintenance, and ventilation and were built large enough for service workers to enter them.[3]

The palace elite weren't the only ones enjoying these Bronze Age amenities. Urban sewer systems that served large city areas existed in several prehistoric Aegean sites, dating between 3400 and 1200 BC. The homes of residents of the ancient city of Thera—located on what today is the island of Santorini—were equipped with toilets and bathrooms directly connected to an underground network of sewer pipes laid beneath the paved streets. The Minoans commonly used ceramic pipes of complex design, shaping the pipe ends so that the pieces could be fit tightly into each other. The pipes' upper parts had openings covered by ceramic lids, allowing for cleaning and maintenance.[4]

The Minoan sewage system was so expertly built that some of it continued to function into the twentieth century. When Italian writer Angelo Mosso visited the archaeological site of another ancient Minoan settlement, Hagia Triada, he was amazed how well the sewer system worked. "One day, after a heavy downpour of rain, I was interested to find that all the drains acted perfectly, and I saw the water flow from the sewers through which a man could walk upright," he wrote in 1907. "I doubt if there is any other instance of a drainage system acting after 4000 years."[5] To which American sanitation and hydraulic engineer Harold Farnsworth Gray later responded, "Perhaps we also may be permitted to doubt whether our modern sewage systems will still be functioning after even one thousand years."[6]

Another sewage system whose remains impress modern sanitation engineers was built by the Harappan civilization that flourished in the Indus Valley. At its height, between 2600 and 1900 BC, the city of Harappa—now an archaeological site near Ahmedabad in the Indian state of Gujarat—counted over 23,000 residents and occupied about

370 acres. Mohenjo-daro was another well-developed city of the Indus Valley civilizations. More than 2,000 years before the Roman Empire would become famous for its feats of engineering, the Harappans built clay brick houses equipped with private toilet facilities that emptied into a sewage structure—a system of covered outside drains.

The Harappans, who lived in a fairly arid climate, had a way with water that was unheard of at that point in history. At a time when the Mesopotamians still carried their water in buckets, the Harappans relied exclusively on wells—vertically built or sunken circular brick-lined shafts—constructed within their urban area. According to some accounts, the city had more than 700 such wells, which afforded its residents the rare luxury of using water not just for drinking and irrigating, but also for cleaning their toilets.

Many Harappan dwellings had private baths that were connected to a sewage system. They used a simple type of drainage in which a bathing room would have a sloping floor so that the water was directed into a specific corner, from which it would flow into an open outside drain or a catchment vessel. The Harappans also built simple latrines—sinkholes over cesspits. More sophisticated ones were constructed in conjunction with the bathing rooms. Those came equipped with seats, made from brick or wood, placed over separate vertical chutes through which the sewage fell, either directly into the drain or into a cesspit.

To route the filth and dirty water out of their homes, Harappan engineers dug 20-inch-deep gutters, which they lined with clay bricks and covered over with wooden boards and loose stones. The covers helped keep filth from escaping the gutters, but could be easily opened at any moment to clean clogged or malfunctioning passageways. The gutters were sloped so that the water could flow, and they joined drains from other houses along the way—much like our sewer pipes do today. Wherever a drain ran a longer distance, or where several drain routes met, the Harappans installed a brick-lined cesspool to avoid overflowing or clogging. Naturally, such cesspools needed to be periodically emptied, so the ancient engineers equipped the cesspool shafts with steps leading down into the pits.

The Harappans and the Minoans were probably the first people who really flushed. They didn't have shiny metal levers attached to gleaming white bowls that sent their excrement into oblivion with a simple button push, but they certainly figured out how to use water to get the filth out of their homes. The city itself probably stunk of decomposing sewage and required frequent cleanings,[7] but despite these setbacks, the Harappans just about solved the pesky problem of their mass-produced dark matter. Unfortunately, although this approach may have worked for Harappa's twenty-something-thousand inhabitants pooping on some 300 acres, or even for 100,000 denizens doing so on a slightly larger piece of land, cities were going to grow much bigger than that.

Only a couple of thousand years later, the city of Rome would hit over a million people—and little street gutters would have choked long before that. Naturally, the sewers would have to be made really big. And that's what the Romans did—they scaled up land-to-water nutrient flushing to grand proportions.

The Romans and the Cloaca Massima

A few years ago, while traveling with my husband in Turkey, I stopped at a few archaeological sites that had belonged to the Roman Empire before the Ottomans took over. On the way from Istanbul to Izmir, we visited Ephesus, a Greco-Roman city that grew to prominence around the second century AD, reaching some 300,000 to 400,000 denizens. After the requisite tour of Ephesus's broad marble-lined streets, the Basilica of St. John, and the Scholastica baths, our guide let us wander off and explore, which we happily did.

Hopping over the fallen columns and blooming red poppies that grow all over Turkey, we suddenly ambled into a large open space, drastically different from anything we had seen before. Straight in front of us was a long white marble bench with a row of holes shaped just like our modern toilet seats. It was a Roman bathroom. A real ancient one.

As I turned around, I discovered two more rows of holes, altogether able to accommodate a small party—or maybe even a small platoon. But those holes were cut so close to one another that I was wondering how people actually used them. Wouldn't they put you in the immediate proximity of someone else's butt? And there were no dividers of any kind in between. Talk about not having inhibitions—doing your private business next to a dozen other straining folks.

Underneath the seats was a stone-lined gutter where flowing water must have carried the citizens' dirty business out of the city. Another, shallower one ran underneath my feet. It, too, was clearly built to carry water—but for what? Other questions brewed in my head. Did the enclosure have a roof, doors, and windows? Were the stone seats hot in summer and cold in winter? Did the toilet-goers talk to each other? Did they shake hands after wiping? And what did they actually wipe with, given that toilet paper is a fairly recent development? Oh, and by the way, was I sitting in a men's room or a ladies' room? Since the doors were long gone, I had no way of telling, had I?

This accidental discovery left such a profound impression that I suddenly found myself obsessed with all these intriguing questions, the answers to which I assumed had long since disappeared into the annals of history—or rather, into its sewers. I was curious whether anyone had ever studied the topic—and sure enough, someone had.

Ann Olga Koloski-Ostrow, an anthropologist and author of a book titled *The Archaeology of Sanitation in Roman Italy: Toilets, Sewers, and Water Systems,* says that her "official" title is the Queen of Latrines. "I live my life in the gutter," she chuckles. "My friends call me Koloski-Ostrow on the toilet. It is an interesting topic to study."

Koloski-Ostrow has lived in that ancient Roman gutter for a quarter century. She had "fallen into it" by chance, when ancient history professor Nicholas Horsfall mouthed the magic words "Roman latrines." Nobody had excavated them the right way, he told her—and she was hooked. "There's a lot you can find out about a culture when you look at how they managed their toilets," says Koloski-Ostrow, who co-directs a program in ancient Greek and Roman studies at Brandeis University in Massachusetts. "That's why I study it."

Over a lovely conversation about bodily excretions, chamber pots, butt-wiping habits, sewer vermin, and other equally unappetizing topics, the ancient Romans' views on waste, hygiene, and toilet habits begin to take shape. One of the first things I learn is that the word "latrine," or *latrina* in Latin, was used to describe a private toilet in someone's home, usually constructed over a cesspit. Public toilets, like the one I tried on for size in Ephesus, were called *foricae*. They were often attached to public baths, the water from which was used to flush down the filth.

The other thing I learn is that because the Roman Empire lasted for 2,000 years and stretched from Africa to the British Isles, Roman toilet attitudes varied geographically and over time. But generally, the Romans had fewer inhibitions than we do today. They were indeed reasonably content sitting in close quarters—after all, Roman theater seats were rather close, too, about 30 centimeters apart. And they were indeed reasonably content taking communal dumps—although the elaborate folds of the toga afforded some seclusion. "Today, you pull down your pants and expose yourself, but when you had your toga wrapped around you, it provided a natural protection," Koloski-Ostrow says. "The clothes they wore would provide a barricade so you actually could do your business in relative privacy, get up, and go. And hopefully your toga wasn't too dirty after that." If you compare the *forica* with the modern urinal, she adds, it actually offers more privacy.

Despite the lack of toilet paper, the toilet-goers did wipe—and that's what the mysterious shallow gutter underneath my feet was for. The Romans cleaned their behinds with sea sponges attached to a stick, and the gutter supplied clean flowing water to dip the sponges in.[8] This soft and gentle tool was called a *tersorium*, which literally meant "a wiping thing." The Romans liked to move their bowels in comfort. Whether they washed their hands after that is another story. Maybe they dipped their fingers into an amphora by the door. Maybe they did not. Maybe they did it in some parts of the empire, but not in others. Worse, the *tersoria* were probably reused and shared by all the fellow butt-wipers who came and went throughout the day. So if one of the *forica* visitors had intestinal worms, all the others would carry them home, too. Without any knowledge of how diseases spread, the overall Roman toilet setup could hardly be called hygienic by our standards.

Even though they looked advanced for an ancient civilization, in reality Roman public toilets were far from glamorous, Koloski-Ostrow says. The white marble seats gleaming in the sun may look beautiful and clean to us today, but that was hardly the case when these facilities were operational. They had low roofs and tiny windows that let in little light. People sometimes missed the holes, so the floors and seats were often soiled, the air stunk, and handwashing was hardly compulsory. "Think about it—how often does someone come and wipe off that marble?" Koloski-Ostrow asks. In fact, she thinks the facilities were so unwelcoming that the Roman elite would use them only under great duress.

The upper-class Romans, who sometimes paid for the *foricae* to be erected, generally wouldn't set foot in these places. They constructed them for the poor and for the slaves who had to run errands for their masters—but not because they took pity on the lower classes. They built them so that they didn't have to walk knee-deep in shit on their streets, getting their togas dirty. Just like any other civilization that chose to settle and urbanize, the Romans were up against a problem—what to do with all this shit? The Roman elite viewed public toilets as an instrument that flushed the filth of the plebes out of their noble sight. Other than that, they wanted nothing to do with the *foricae*. In a Roman bath, it was common practice to inscribe the name of the benefactor who paid to build it, but toilet walls bear no such writing. "It seems that no one in Rome wanted to be associated with a toilet," Koloski-Ostrow says.[9]

After all, why would the refined noblemen want to sit next to the common people who had lice, open wounds, skin sores, diarrhea, and other health problems? And that wasn't the worst of it. The sewers underneath were a welcoming home for vermin—like any other sewers anywhere in the world. "Rats, snakes, and spiders would come up from down below," Koloski-Ostrow adds to the already gruesome picture. Plus, the decomposing sewage may have produced methane, which could ignite—quite literally lighting a fire under someone's ass. "With all that, I concluded that the Roman public toilets were probably not used by the Roman elite," she says.

Neither were the public toilets built to accommodate women. "By the second century CE, I don't think women used them," Koloski-Ostrow says. "It was mostly the men's world. The public latrines were constructed in the areas of the city where men had business to do. Maybe a slave girl who was sent to the market would venture in, out of necessity, although she would fear being mugged or raped. But an elite Roman woman wouldn't be caught dead in there."

Back at their comfortable villas, the wealthy citizens had their own personal latrines constructed over cesspools, but even they may have preferred to use the more comfortable and less smelly chamber pots, which they let the slaves empty onto the garden patches. They didn't even want to connect their cesspools to the sewer pipes because that would be likely to bring the vermin and the stink into the house. Instead, they hired *stercorraii*—manure removers—to empty their pits. Koloski-Ostrow writes that in one case, "11 asses may have been paid for the removal of manure."[10]

The famous Roman *sewers*, however, were another story. At the height of its power, the city of Rome had to clean up after about a million people, so the small, 20-inch-deep, wood-covered gutters of the Minoans and the Harappans wouldn't do. Rome had 10 times more inhabitants than Knossos did at its height, and thus produced 10 times more waste, totaling 500 tons a day. And while Roman farmers understood the waste's fertilizing value and put some of it back into the fields, the city just couldn't recycle it fast enough. A half-million-ton pile of shit is a truly mind-boggling image. To flush that much excrement out of the city daily, one needs a truly massive system. The Romans, as my husband aptly pointed out, did everything on a grand scale—including filth removal.

The Romans gleaned their sewer technology from the Greeks. In her book, Koloski-Ostrow attributes that "technology transfer" to "Hellenistic cultural forces" and the Roman soldiers who starting building latrines in their military camps. "By the first century B.C., and probably much earlier, the cities of Roman Italy had already learned latrine technology well from the Hellenistic East," she writes, "and transformed it for their own Roman urban planning needs." But to keep their Roman-sized Augean stables clean, they scaled up their system to massive proportions. They built the Greatest Sewer, or Cloaca Massima, named after the Roman goddess Cloacina—the Cleanser, from the Latin verb *cluo*, meaning "to clean."[11]

The Cloaca Massima moved millions of gallons of water and flushed about a million pounds of crap a day. It was so immense that Greek geographer and historian Strabo wrote that Roman sewers were big enough "for wagons loaded with hay to pass" and for "veritable rivers" to flow through them. He wasn't that far off in his descriptions. The Cloaca Massima accomplished several things: it drained the excess water from the city, rid the people of their waste, and generally carried away everything they didn't want, discharging it all into the River Tiber. It also drained water from the surrounding swamps and river valleys, preventing floods. Roman author and naturalist Pliny the Elder wrote that when the rivers surrounding Rome spilled into the sewers with unrelenting force, the sewers withstood Mother Nature's wrath, directing the currents down to the Tiber, where the triple-arch outlet of the Cloaca Massima still stands today.[12] Another Greek historian, Dionysius of Halicarnassus, measured the greatness of the Roman Empire by its sewers, as well as by its aqueducts and paved roads. When the sewers clogged up or needed other repairs, a considerable amount of money was spent on keeping them functioning. Despite many earthquakes, floods, collapsed buildings, and other cataclysms, the Roman sewers stood strong over centuries.[13]

The Cloaca Massima solved Rome's sewage removal problems, but it didn't solve the city's health issues. It carried the filth out of the city and dumped it into the Tiber, polluting the very water some citizens depended on for irrigation, bathing, and drinking. And so, while the Romans no longer had to see, or smell, their excrement, they hadn't done much to eliminate its hazardous nature. Through the next several centuries, as humankind kept concentrating in cities, it would find itself in a bitter battle with its own waste, seemingly with no way to win.

And yet, some societies were remarkable exceptions to this never-ending fight against their excrement. They not only perfected their waste recycling, but treated their waste as delicately as they would handle "honey from the hives," and measured its value in gold.

CHAPTER 3

TREASURE NIGHT SOIL AS IF IT WERE GOLD!

At some time during the Tokugawa period in Japan, the magistrates in the coastal city of Osaka received an air quality complaint. Residents who lived near the port objected to the foul smell that emanated from some of the docking ships. A thriving urban sprawl, Osaka welcomed a number of merchant boats, domestic and foreign, that delivered tea, rice, silk, fish, and other goods. But along with these boats came other vessels that ferried much less agreeable cargo—namely, human waste.

These boats cleared the city of its daily *shimogoe*, as the Japanese called their sewage.

Brought to the wharves by the *shimogoe* collectors, the waste was loaded into the boats' bellies. But instead of being dumped into the sea or onto a remote island, where the Japanese sent some of their other, less useful garbage, it was shipped to local farmers. The valuable *shimogoe* went on to nourish food for the people who produced it. Fittingly, the word *shimogoe* literally meant "fertilizer from the bottom of a person," and is roughly translated into English as "night soil," explains Kayo Tajima, professor at the Rikkyo University in Tokyo, whose research focuses on environmental economics and urban studies.[1] As the city grew, so did the amount of night soil it generated. More boats had to come to take away the mammoth loads. Eventually, the daily hauls generated such a stink that people protested.

The magistrates considered the problem. On one hand, the complaint had merit. The Japanese culture valued cleanliness. On the other, banning the sewage ships from the port would cause not one, but two major problems. First, it would cripple the city's waste disposal system. Second, it would leave farmers without fertilizer and urbanites without food, both of which would result in an uproar. After much deliberation, the magistrates ruled that "it was unavoidable for the manure boats to come into the wharves used by the tea and other ships." In the end, the sewage haulers retained their rights to dock alongside other vessels.[2]

To us, this decision may seem unhygienic at best. But to the Japanese, the ruling was logical. They maintained a very different view of human excreta. Unlike European countries, Japan was not blessed with an abundance of natural resources and large swaths of fertile land. As a little country spread over a few small islands where mountain ranges occupied three-quarters of the landmass, Japan had to make do with what it had. A large portion of its harsh, rocky terrain couldn't be used for agriculture at all. It didn't have the abundant grassy landscapes so common in Europe, which limited how many cattle it could sustain. Its soil, sandy and nutrient-poor, would bear very meager crops without fertilizer.[3] While a European farmer could count on getting a rich harvest from a freshly cleared patch of forestland, the Japanese never expected much from a new plot. A Japanese saying, "A new field gives but a small crop," epitomizes the agricultural challenges the culture faced.

As crops grow, they extract nutrients from the soil—including nitrogen, carbon, calcium, phosphorus, potassium, sulfur, magnesium, and others—to build their cell walls. If the soil is to produce crops year after year, these nutrients must be replenished—in the form of organic matter regularly returned to the ground. This organic matter can be agricultural refuse like rice husks, animal remains like crushed bones, or digested food like manure—which contains all of the above nutrients from the eaten plants.

Once in the soil, all of that becomes microbial food. When it comes to breaking down and decomposing organic materials, different microorganisms have different jobs. Caitlin Hodges, who studies soil science at Pennsylvania State University, breaks them into three distinct groups: recyclers, miners, and refiners. Recyclers, such as fungi and bacteria, break down plant and animal matter. They feed on the carbon in that organic refuse, using it as an energy source, and they free nutrients like nitrogen, potassium, and phosphorus to be taken up by the next generation of plants. Miners produce bacterial exudate—a form of chemical goo—that extracts nutrients like phosphorus, calcium, and potassium from rocks and minerals in a plant-ready form. Refiners are the nitrogen-fixing organisms. Called rhizobia, they are a unique type of bacteria. Millions of years before humans learned to make synthetic nitrogen, the rhizobia evolved to convert the highly inert nitrogen gas molecules in the atmosphere (N_2) into ammonia molecules (NH_3)—a potent plant food. Breaking apart the nitrogen molecule, which is very stable thanks to its triple atomic bond, is a tough job that requires a lot of energy. The bacteria accomplish it within their own cells using a special enzyme they evolved, called nitrogenase. Mary Stromberger, a soil microbiologist at Colorado State University, finds this ability fascinating. "Humans use very high temperature and pressure to do it," she says. "But these bacteria do it within their own cells at room temperature and atmospheric pressure." So far, humans haven't come close to such perfection—in part because the nitrogenase enzyme is very sensitive to oxygen, Stromberger says. The microbes had probably evolved it before the earth's atmosphere became oxygen-rich.

Converting atmospheric nitrogen into plant food is a very slow process. Freeing nitrogen and other nutrients from agricultural refuse is somewhat quicker, yet it still takes time—weeks for leaves and months for twigs or cornhusks. But manure, whether human or animal, is a partially digested organic material, so it's already broken into smaller compounds and molecules, and that makes the microbes' work easier. Composted manure, which microorganisms and earthworms have already chewed through, makes an even better fertilizer. The plants don't have to wait for the nutrients to become available—their roots can just start sucking them all up.

Japanese farmers didn't know about the complex inner workings of the soil microbes, but they could certainly see that adding humanure to their land made crops flourish. They didn't have a lot of domestic animals due to their lack of pastureland, so the primary fertilizer they counted on was the one they collected from themselves. "Japanese peasants didn't keep big animals, they had very few horses and cows, so they didn't have animal manure," Tajima explains. "So they had to use what humans produced—*shimogoe*."

German botanist and traveler Philipp Franz von Siebold wrote about how impressed he was with the Japanese ability to convert their barren, infertile soil into flourishing fields. "The soil is naturally sterile, but the labour bestowed upon it, aided by judicious irrigation, and all the manure that can in any way be collected, conquers its natural defects, and is repaid by abundant harvests."[4]

So it's not surprising that for Japanese farmers, excrement was a prized resource on which they placed a high economic value. It was certainly worth the stink. The collectors, who went from door to door to gather the riches, paid a pretty penny for it. The haughty Osaka urbanites offended by the smell could not win against the treasured waste. That humanure was needed to feed humanity.

And humanity was rapidly growing.

The Excrement Economy

Even before the Tokugawa period, which started in the seventeenth century, Japan was undergoing a massive urban boom. The little fishing villages that dotted the country's shores began to grow into towns, and towns swelled into cities. By the early eighteenth century, Edo, which was later renamed Tokyo, was larger than any European city at the time: London's population was estimated at about 575,000, and Edo's at about a million. All of Edo's urban dwellers relied on the surrounding farms to supply them with rice and vegetables. And the farmers relied on the cities' waste to grow their produce year after year.

Farms that were relatively close to the city's outskirts could collect *shimogoe* without the expensive boat transport. They would mount two buckets, called *koe-oke*, on the ends of a pole—just like my grandfather did with a *koromyslo*—and carry it to the farm on their backs or with a pushcart. That load was called one *ka*. A horse could carry four *koe-oke* or two *ka*. Some farmers came to town to collect the fertilizer themselves. Those who could hire help sent day laborers. Peasants who lived farther away and couldn't afford to buy a boat had to rely on rich farmers' willingness to sell them some *shimogoe*. And those able to buy a boat could start a lucrative *shimogoe* shipping business.[5]

In some cases, farmers established annual contracts with urban families. Called *tsuke-tsubo*, these agreements stipulated that a farmer could collect all of the household's night soil for a year in exchange for a certain amount of rice as a down payment. So the residents would save their *shimogoe* for their particular farmer. Some upper-class farmers had established relationships with *daimyô*—Japan's feudal lords who governed the country's provinces and owned large estates with many servants. The farmers supplied the estates with firewood and young plants for vegetable gardens, in exchange for the privilege of emptying the estates' toilets. The access to such riches was a boon, not only in quantity, but also in quality, Tajima explains. The nobility and their servants ate well, so they produced nutrient-rich *shimogoe*.

Deemed an important business, *shimogoe* transactions were handled in a very businesslike manner—through regulations, contracts, and records. "Edo/Tokyo's highly commercialized system for managing night soil was to a large extent unique," Tajima writes in her paper titled "The Marketing of Urban Human Waste in the Early Modern Edo/Tokyo Metropolitan Area," which describes the sophistication and structure of city-farm night soil relationships.[6] There was also a process for resolving the conflicts that periodically arose over waste. The clashes weren't about whose unpleasant duty it was to remove the muck, but rather about who would be lucky enough to lay their hands on it.

Japanologist Susan B. Hanley also notes that "the most important difference between waste disposal in Japan and in the West was that human excreta were not regarded as something that one paid to have removed, but rather as a product with a positive economic value."[7] In fact, the value of this waste wasn't just positive—it rose as the cities grew and fluctuated with the seasons. Tajima cites historical sources that chronicled what farmers of the Tokugawa period were willing to pay Edo's residents for their *shimogoe*. "Seven or eight years ago they used to exchange the urine for rice straw; in winter until the beginning of spring, they gave two small bundles of straw for one *ka* (two buckets) of urine. From the end of the second month to the third month in spring, they gave three to four bundles of straw for the same amount of urine, and in the beginning of summer, when a lot of fertilizer is needed, they gave six bundles of straw."[8]

Similar "waste inflation" was happening in eighteenth-century Osaka. Before then, Osaka's residents simply exchanged their excrement for food. The incoming boats would bring in produce and fish, and the outgoing ones would carry excrement off to the countryside. But in the early eighteenth century, food demands grew. More rice paddies were built in the countryside around Osaka, and they all needed to be fertilized. Food prices jumped as well, and the urbanites, no longer able to afford bartering, started selling their excrement to the farmers. It wasn't cheap. The price of the night soil produced by 10 households in a year was about two to three *bu* of silver or over half a *ryo* of gold. To put that in perspective, one *ryo* could buy all the grain needed to feed one person for a year.[9]

A system of regulations governed the night soil business, delving into minuscule details unfathomable by our standards. For example, if a family rented a house, who had the rights to the excrement— the tenants or the landlord? It may seem logical that the tenants, who produced it, should've been the proud owners of their poo, but that's not how things worked in frugal, nourishment-poor Japan. The dwelling's night soil belonged to the landlord, who sold it, by contract, to the night soil gatherers. What's more, the price paid was factored into the rent: as Hanley writes, "Rent was adjusted on the basis of how many tenants there were and was raised if the number of occupants dropped." American zoologist Edward Morse, who lived in Japan, wrote that he "was told in Hiroshima in the renting of the poorer tenement houses, if three persons occupied a room together the sewage paid the rent of one, and if five occupied the same room no rent was charged." Human beings and their natural power to produce this sustainable and valuable resource were held in great esteem.[10]

European travelers to Japan told stories of farmers bringing gifts to community members near Tokyo to thank them for their manurial donations. "Once a year, in fact, the terrace people were given what they called 'dung cakes.' A local farmer used regularly to bring along a cart and buy up all the night soil from the communal toilet; then, at the end of the year, he'd take them some of the special rice used for making rice cakes to thank them for the year's supply of 'dung' . . . 'you can see 'business' has been good this year—there are plenty of dung cakes,' we'd joke to each other."[11]

Farmers fought over who got to collect excrement and where. In the summer of 1724, two groups of villages from the Yamazaki and Takatsuki areas erupted in "poop wars," fighting over the rights to gather night soil from different parts of Osaka. So did farmers from other areas. The cities formed their own organizations charged with organizing and overseeing waste disposal, trading, and price negotiation. These city guilds argued with farmers' associations over monopolies on night soil collection, areas, and fees. As the city guilds raised the price, the less fortunate farmers who could no longer afford to pay for their fertilizer were in trouble. That led to a crime unfathomable by Western standards: stealing shit. Incidents of such theft appear in the Japanese records more than once. By Japanese canons, it was a very serious offense. So serious, in fact, that the law-enforcement authorities punished it by sending the felons to prison. And yet that didn't stop the desperate farmers from committing their stinky crime. A sudden drop in excrement supply could completely devastate a family. When a fire burned down a large residential area in Edo that one farmer had relied on, he "suffered major crop losses."[12]

Urine, which actually contains more nitrogen than feces, was also in demand. In many cases, the two were collected and used as a mixture,[13] but in some areas they were collected in separate pots and marketed separately. Urine, however, was more problematic to ship, in part because it was a liquid and in part because humans produce a much greater volume of it than the other thing. Solid waste was more compact, harder to spill, and easier to transport over long distances. But the smart and frugal village folk didn't let any accidental nourishment slip away. If a traveler passing through the village had to use the facilities, they were readily provided in the form of buckets left by the side of the road so that not a precious drop would go to waste. Some villagers even built privy vaults on the street-facing side of their homes for the convenience of passers-by—another feature that left a deep impression on European travelers. "Care is taken, that the filth of travellers be not lost, and there are in several places, near country people's houses, or in their fields, houses of office built for them to do their needs," German explorer Engelbert Kaempfer wrote in his book about his journeys in Japan. "Nor doth horses dung lie long upon the ground but it is soon taken up by poor country children and serves to manure the fields."[14]

The excrement producers understood how much power they carried within them. And they could be quite discerning about where to exercise that power—at their home or at somebody else's. Some could be generous. They would take a dump at their in-laws' house while visiting for dinner. Others could be tacky and miserly, and keep it all to themselves. That behavior could apparently earn you a bad reputation. The Japanese "told stories about stingy guests who would hurry home when they felt their sphincters tightening so as not to give away valuable fertilizer."[15] Going to the bathroom at a friend's house was an act of generosity. It was like leaving a gift.

How did the Japanese avoid the deadly diseases that come with raw sewage? First, it was probably because farmers practiced composting. They would unload the manure, mix it with other organic waste, and let it sit for a while, just like my grandfather did, rather than immediately spreading it onto the crops. Bacterial decomposition would run its course, breaking waste down into plant food— nitrogen, phosphorus, potassium, and other elements. These basic building blocks of life appeal to plants, but less so to pathogens, which prefer the protein- and carbohydrate-rich food found in fresher, more recent manure. (If you've noticed, flies immediately gather over a pile of dog poop, but rarely crowd over plain dirt. Compost isn't their favorite dinner, either.) Second, the Japanese hygienic culture helped. Long before science learned about germs, the Japanese diligently washed their hands and bodies and boiled their water to make tea. Historians also write that the Japanese consumed a lot of their vegetables cooked rather than raw, which helped kill any pathogens that survived in compost and made it onto the produce. Whether the Japanese arrived at all that wisdom by trial and error, or simply by chance, it worked. They never suffered from infectious disease epidemics to the extent Western societies did. The plagues that swept through preindustrial Europe, killing hundreds of thousands of people, spared the Land of the Rising Sun.[16]

But the Japanese weren't the only successful excrement recyclers in human history. The Chinese may have started it even earlier. And, given their much more populous cities, they operated their lucrative recycling businesses on a much grander scale.

A Precious Gem

"Treasure Night Soil as if It Were Gold!" was the directive of an imperial treatise published during the early Qing dynasty in 1737. Written after a comparison of the agricultural practices of various geographic regions within China, the document was a directive to the northern farmers who hadn't yet fully caught up with the manuring wisdom of the southerners. Unlike the more progressive residents of Jiangnan Province, the northerners weren't keen on collecting their waste and were ignorant of its value. "Therefore, the streets in the north are not clean. The land is filthy," the treatise read. "The northerners should follow Jiangnan's example." The empire's final verdict was plain and simple: "Every household should collect night soil."[17] End of discussion. Try issuing a verdict like that in our industrial world today. We still can't consistently recycle plastic bottles, let alone something as unruly and unpalatable as our own muck.

There was a reason why the southerners were more adept at the night soil business. Hangzhou, located on Hangzhou Bay in the southern part of the county, was a major seaport and one of the largest cities on earth. By the 1820s, its population exceeded 3 million people. The municipality of ancient Rome had produced a smaller daily excrement pile than Hangzhou residents did. And all the residents of this urban sprawl had to be fed. So the Hangzhou farmers and urbanites forged an excrement economy similar to those of Edo and Osaka, only three times the size. Another massive city, Suzhou, located on the Yangtze River delta, moved around twice as much shit. In 1851, it cleaned up after 6.5 million inhabitants. Nanjing, also on the Yangtze River delta, gathered and shipped the night soil of 6.2 million residents. That's about as many people as the city of Rome has *today*.

Called *fenfu*, the night soil middlemen wheeled their carts through the city streets, collecting waste from homes and grounds in wooden containers, each of which could hold about 60 pounds. Carts could fit six to ten of them, accumulating up to 600 pounds total. Those who didn't have carts started their businesses by carrying buckets on poles thrown over their shoulders. The Chinese may have actually coined the term "night soil." The name derives from the fact that the fertile muck was gathered in the early mornings when the residents put their chamber pots out the door. The operation was so profitable that it lasted into the twentieth century. Neither the waste nor its collectors were stigmatized, and in a way, that benevolent attitude still remains today. Journalist Rose George, in her book *The Big Necessity: The Unmentionable World of Human Waste and Why It Matters*, remarked that "of all the peoples of the world, the Chinese are probably the most at home with their excrement."[18] It was part of the cycle of putting food on the table—just like it was for my grandfather and me. Only instead of hiring professional *fenfu*, he did it himself.

The *fenfu* were efficient. They figured out their daily routes, the best way to transport the muck out of the city, and where to find buyers. They sailed it out to the country in gondolas, covered with straw and washed on the outside to keep out the stink. They brought the dung to large empty lots outside the city and processed it—spread, dried, and sorted it. Not all shit was created equal. The offerings of Hangzhou's elite sold to the highest bidder because they ate better and had more diverse diets. The poor people's dung earned less. Pig and sheep manure sold, too, but not for as much as humanure, and was good for only certain types of crops. And urine had a price tag, too. Were there counterfeit manure schemes in which pig dung masqueraded as Hangzhou's premium product? Who knows? One would think that farmers knew their shit, but perhaps some old court documents in Han Chinese can speak about various excremental disputes. As in Japan, excrement theft was a real threat. "So precious was manure the Chinese farmers stored it in burglarproof containers," according to Gene Logsdon, a "contrary farmer" who has published several books on agricultural subjects. Just as it was in the Japanese culture, gifting excrement was a gracious gesture. "The polite thing to do after enjoying a meal at a friend's house was to go to the bathroom before you departed."[19]

Historians aren't sure exactly when Chinese farmers learned to fertilize crops with humanure. Some date this spark of genius to 2,000 years ago, others to 4,000 years ago, but regardless of the exact timing, China is one of the oldest continuously functioning agricultural societies on earth, writes Donald Worster, an environmental historian, in his treatise "The Good Muck: Toward an Excremental History of China." Many societies failed to farm sustainably over centuries, but the Han people, who became the dominant ethnic group in China, were more successful than the Fertile Crescent folks. Worster writes that the Chinese may have noticed that rice paddies, which need to be frequently fertilized, flourished after human waste was tossed into the mix.[20]

American agricultural scientist Franklin Hiram King, who had once been chief of the Division of Soil Management in the USDA Bureau of Soils, was equally impressed by the Chinese peasants' ability to maintain their soil fertility—something that American farmers hadn't succeeded in doing. Concerned about the rapid deterioration of American soil, King traveled to Asia in 1909 to learn its "permanent agriculture" methods, which, in his book *Farmers of Forty Centuries*, he suggested American farmers should adopt. His suggestions didn't sprout strong roots at the time, but modern organic farmers like Logsdon cited the same ideas: "In Japan, Korea and China manure was treated like a precious gem because it was a precious gem. Every scrap of animal waste, human waste and plant residue was scrupulously collected and reapplied to the land."[21]

German chemist Justus von Liebig,[22] who studied how plants extract minerals and nutrients from the soil, and was the first to realize the importance of nitrogen for agriculture, was also very impressed with Chinese farmers. He wrote about how they managed to utilize all kinds and forms of waste by setting up "tubs, sunk into the ground" or constructing "large cisterns and pits lined with lime plaster," in which they would toss all kinds of human and animal waste along with agricultural refuse. They added water, covered the tubs and pits with straw—to keep the moisture and the stink sealed in—and left it all to "undergo the putrefactive fermentation." When nature had taken its course, they would pour this liquid manure onto the roots of plants and fruit trees. Those who didn't farm would shape their excrement into "cakes," dry them in the sun, and sell the fertilizer to their countrymen. The latter would mix the cakes with water and take the liquid out to the fields.

To use the dung in the most efficient way, farmers applied it to the roots of their plants, rather than just spreading it over the fields. Liebig writes that except for rice paddies, the Chinese always applied manure to "the plant itself rather than the soil, supplying it copiously with their liquid preparation." They even steeped their seeds in liquid manure before planting until the seeds swelled and popped, because that nutrient-rich medium would speed up the germination process and produce healthier plants. Animal dung was used, too, but night soil was "esteemed above all others." Urine was not wasted either. "Every farm, or patch of land for cultivation, has a tank, where all substances convertible into manure are carefully deposited, the whole made liquid by adding urine in the proportion required, and invariably applied in that state."

European farmers didn't bother with all that dirty business. Liebig, who was one of the few people who actually understood the biochemical cycle of nutrients in soil, lamented his countrymen's ignorance as he praised the Asian methods: "The Chinese are the most admirable gardeners and trainers of plants, for each of which they understand how to prepare and apply the best-adapted manure." To Liebig's chagrin, this wasn't a widespread practice in his own country. "How infinitely inferior is the agriculture of Europe to that of the Chinese!" he wrote. "The agriculture of their country is the most perfect in the world."

Tout-à-la-Rue

Contrary to East Asians, medieval Europeans had a very negative opinion of their waste. They had several reasons to be dismissive about the potent substance they all produced regularly. First, they had enough rich soil to bear sufficient crops. Second, they herded cattle, whose manure was naturally dispersed in the fields, replenishing some of the nutrients. And so, they could afford to view their waste as a burden, a nuisance, disgusting filth that had to be removed and destroyed.

In "Manure and the Medieval Social Order," historian Richard Jones explains that while the medieval farmers regularly spread cattle dung on their fields, they tended to shun using human waste for their crops. "In medieval England few issues were as vital as soil fertility and few substances more important than manure," he writes. Yet even peasants, who, unlike the lords, had few animals, and thus much less fertilizer to replenish their fields, didn't embrace their own riches. According to Jones, fertilizing guidance advised "estate bailiffs to spread surplus straw and bracken onto muddy ground or roads in order to make compost" and to add "straw and litter from cow sheds and sheepcotes." Yet there was not a mention of putting their own waste to good use—contrary to the Qing dynasty's directive to "treasure night soil as if it were gold." Jones's conclusion, based on historical sources, is that "indeed the human contribution to the manorial manure heap appears to have been taboo."[23]

And thus, with few exceptions, European urban waste management was focused not on sending waste to farmers, but on sending it as far away as possible.

During the medieval age, the poor built outdoor latrines, sometimes encasing the waste in barrels to contain it. The wealthy, having gleaned some technology from the Romans, built toilet rooms within their castles, which they called garderobes. Equipped with stone or wooden seats with holes underneath, these facilities were built over the moats that encircled the castles. The excrement would fall down a chute and into the water, where the currents would eventually carry the unwanted matter away, but not before the smells rose up and wafted back in through the windows. To do their business, the nobles had to rid themselves of their elaborate robes, which they often left for their servants to hold or guard—hence the name "garderobes."

These waste disposal methods worked as long as the castles were relatively isolated and had reasonably few people living in them, but when Europeans began to aggregate in cities, things got ugly. In the crowded tenements of medieval London there was a shortage of latrines, so the waste was simply thrown into the alleyways. Berlin residents piled up their shit in front of St. Peter's Church until a 1671 law obligated every visiting peasant to take a load home. But the medieval Parisians may have been the worst offenders. A law passed in 1531 required landlords to provide a latrine for every house, but it wasn't enforced, so the poor defecated any place they saw fit.[24] The upper-floor residents accumulated their waste in chamber pots, placed in night "commodes"—bedside cabinets with doors to keep them out of sight. Some of the commodes were elegant pieces of furniture, but the disposal of their contents was anything but elegant. The chamber pots were often emptied out the window in a method infamously dubbed by the French as *Tout-à-la-Rue*—"all onto the street." Unsuspecting passers-by would be warned by the cry *"Gardez-l'eau!"*—which meant "Watch out for water!"—to take quick evasive action. This phrase may have led to the English nickname "loo" for the lavatory. The entire city of Paris smelled like one, too. French philosopher Michel Eyquem de Montaigne complained that it was impossible to rent a place in the city where the air didn't reek.

The French royalty wasn't much better. "The Louvre was a mess," writes Gray, who compared the sanitary habits of the Bronze Age and medieval Europe with our own. "People defecated without restraint or attempt at secrecy in the courtyards, on the stairs and balconies, and behind doors, without hindrance from palace attendants. On August 8, 1606, an order was given prohibiting any resident of the palace of St. Germain from committing a nuisance therein. That same day the king's son urinated against the wall of his room."[25]

The Louvre excrement battles were nothing short of amazing. "A favorite locality was the terrace of Tuileries, which eventually became so fouled that superintendent of the royal grounds installed a latrine, charging an admission fee of two sous." That did not go over well. "Enraged at the high price, the public removed their excretory affections to the Royal Palace grounds." To rectify that, the Duc d'Orleans constructed a dozen privies. It's not clear whether they were free of charge or not, but apparently he was more successful in keeping "the nuisance" out of sight.[26]

These descriptions may make it sound as if all of Europe was knee-deep in dung, but that wasn't necessarily true. At times, some cities managed their waste better than others. Some European communities also recognized that excrement was valuable and organized night soil collection processes similar to those of Asian cultures. Collectors, dubbed "scavengers" or "night men," made their gathering rounds at night, scooping out the contents of privy vaults and emptying chamber pots left outdoors for pickup. Sometimes they carted their malodorous harvest to the countryside, and sometimes they sold it to farmers who came to the city to sell their produce. Some night men printed elaborate cards that described the services provided and the benefits conferred. One collector promised that he "decently performs what he undertakes. Empties Vaults and Sespools, unstops Funnels and Cleans Drains, at the very Lowest Prices. N.B. I have the new invented Machine Cart for the Quick dispatch of Business."[27]

Some night men went to great lengths to turn the dispatch of "business" into a working business indeed. They accumulated it in pits and even used some chemicals to bring down the smell. Unlike the Chinese, the Londoners couldn't count on the sun to bake their poo into cakes, so they dried it by more extensive means. Liebig wrote that "owing to the variable nature of the climate, artificial means are employed in its desiccation." The night soil was "placed upon iron plates heated by means of furnaces." In some big cities in the United States, manure manufacturers took the night soil "whilst still soft" and "mixed it with the ashes of wood, or with earth, or both," which also seemed to alleviate the stink.

None of these methods, however, grew into the robust recycling systems typical of Asian waste management—with one notable exception, those of the Flemish and the Dutch.

Like Honey from the Hives

Every spring and autumn, when it was time to plow the land and sow the seeds, Flemish farmers would wheel out fertilizer carts from their barns. Equipped with large casks capable of holding more than 100 gallons of fluid, the containers looked like the ordinary water carts of the time, but they dispensed a very different type of stuff—namely, liquid manure. The casks had bottom valves that could be opened by pulling a rope. As the farmer rode on the horse that pulled the cart through the field, he would tug at the rope to dispense the fertilizer onto his land.

This practice was so superior to most European agricultural methods of the nineteenth century that it caught the attention of European journeymen and writers and resulted in a book, titled *The Outlines of Flemish Husbandry*, describing the locals' talent for making crops flourish despite the odds. Travelers from other European locales were astonished by the diligence with which Flemish and Dutch farmers gathered and preserved their cities' night soil and then shipped it to the countryside in boats and barges. Liebig wrote that the business of collecting urine and night soil employed "an immense number of persons, who deposit tubs in every house in the cities for the reception of the urine of the inmates, which vessels are removed daily, with as much care as our farmers remove their honey from the hives."[28]

Flanders and the Netherlands were two places that approached the business of human waste in like manner to the Japanese. What made them such remarkable outliers? The answer is simple: they had to survive. Like those in Japan, their soils were "light," meaning naturally poor in nutrients, and their winters were longer and colder than those in England. In those conditions, no fertilizer could be spared.

Faced with the stark reality of their lands, the Flemish threw away nothing. Urban solid and liquid wastes were sold to collectors, sometimes together, sometimes separately, and families generally kept tubs or tanks sunk into the ground where they collected household refuse and every bit of organic rubbish. Wasting waste meant losing money. In Ghent, for example, the servants enjoyed a special "perk": they were paid for the bodily liquids they collected, often as much as they received in wages (although some argued that it brought their wages down instead). The observers were impressed by how diligently the Flemish handled their refuse. "The thrifty housewife and her active substitute the maid know the value of what in our households is thrown away and lost," they wrote in *The Outlines of Flemish Husbandry*. Everything was "carefully kept in this reservoir until once a week a farmer or contractor calls with his tub on a cart; and this mixed with contents of privies, which are frequently emptied, he keeps in large cisterns for use and sale." The Europeans were equally surprised by how clean and dung-free the Flemish city streets were. That was because every piece of dung was either swept up or scooped up to add to the precious loot.

Flemish waste dealers dug deep tanks near rivers and canals, from which barges would pick up urban sewage regularly. Lined with bricks and cemented over, the tanks were built to prevent the effluent from seeping out, and their contents were sold to farmers, retail and wholesale. Over summers and winters, the tanks would slowly fill up with the never-ending supply—and then would sell out. "When the season arrives for sowing, in spring and autumn, the farmers come with their carts and tubs and purchase as much as they want."[29]

From that carefully collected organic fortune, Flemish farmers made all kinds of manuring recipes and concoctions, which they applied to their barren lands at certain times of year and at various soil depths—sometimes pouring it over the surface and sometimes working it in by tilling the earth. In addition to sprinkling liquid manure over the fields with watering carts, they distributed dry dung onto patches in small heaps and poured urine into it "to excite fermentation." They also set up compost piles, where they mixed together organic matter and dry manure, poured in some urine or water, and left it to ferment, much as my grandfather did with his compost pits.

Those piles had a special name: they were called *croupissoir* in French and *smoor hoop* in Flemish, which meant "a smoldering heap," because the decomposing mix generated heat.

Like the Chinese *fenfu*, the Flemish differentiated among manure types and applied them to different cultivars. Cow manure was valued very highly and used for dry, sandy lands. Horseshit came in a close second, helping to fertilize clay soils, while pigsty loads were considered less desirable because pigs ate garbage, so their "dung was poor." The chickens, ducks, and geese, as well as pigeons, that left droppings everywhere helped enrich the "kitchen gardens." The "thrifty housewives" swept up all of that good fortune and used it to grow their soup vegetables. And shepherds would gather the dung left by their flocks in the meadows and sell it.

Not a drop of manure was wasted—just as a drop of honey wouldn't be lost either.

English dung drying and Flemish manure sorting were right on par with the *fenfu*'s recycling methods, but they never fully took hold on the European continent. The root cause of this failure was in the economics. Most European urbanites didn't *sell* their excrement in order to bring food to their dinner tables the next year. They *paid* to have the filth removed. And even when the European scavengers sold their product to farmers, the latter had no pressing need to buy it. Most European farmers had domestic animals whose manure they used instead. And when their old fields grew barren, they would simply clear more land. My grandfather had no more land to clear, so he had to get thrifty instead. That's why we also treated our cesspool contents as if it were gold. Or honey from the hives.

Excrement recycling took hold in societies that were short on fertilizing resources and learned not to waste their waste early on. Whether they regarded it as honey or treasured it like gold, they needed it to survive. Most of Europe didn't. So shuttling excrement between cities and villages has never become a lucrative business. And when Europe's cities grew in size, they created an even bigger gap between the urban excrement makers and the countryside excrement users. The larger the cities swelled, the farther away were the fields. Carting human excrement from the growing urban sprawls over the ever longer distances to the farms was becoming an increasingly annoying chore that no one wanted to do.

Archaeology professor Amy Bogaard notes that the problem of waste has persisted since Neolithic times. In her paper "Middening and Manuring in Neolithic Europe,"[30] she shows that manure or midden—domestic waste consisting of kitchen scraps and human excrement—were always a pain to haul, with few willing takers. "Being heavy to transport, midden material and manure are generally spread within a limited zone adjacent to settlement or stalling area," Bogaard notes. Throughout history, the Japanese and Chinese *fenfu* and the Flemish farmers were among the few willing transporters of the substance.

But during Neolithic times, there were fewer people, much less waste, and no big cities at all. That was no longer the case in medieval Europe, where cities were generating sewage at faster rates than it could be used. When scavengers couldn't find buyers for their product, they would skip the long ride to the countryside and simply dump the waste in gutters and rivers, contributing to the problem rather then solving it. Unavoidably, the shit began to pile up again. And the only way to clean the massive Augean stables of the burgeoning European cities was by diverting the force of water into them.

Enter the Water Closet

Since urban Europeans didn't put their waste to good use, their efforts were focused on finding the most efficient ways to eliminate it from their homes. A major breakthrough in these efforts was achieved not by an engineer or a scientist, but rather by a poet—albeit not a very good one, according to historians.

It was made by Sir John Harington, Queen Elizabeth's mischievous cousin, who kept falling in and out of Her Majesty's favor for his not-so-polite and risqué verses, among other things. Sent into exile in Bath by the mighty queen in 1584,[31] Harington busied himself with various construction projects, including the first modern flushing toilet—at least the first since the Minoan and Harappan version.

Harington's toilet was by all means a more advanced version. It used a system of handles to pour in water from a cistern while simultaneously opening the valve levers to flush the contents down the pipes. And then voilà! The unwanted yucky deposits would spiral down the drain—out of sight and out of mind. Where the pipes led was, of course, another story. They didn't lead very far, usually to a water source nearby or a cesspool. Or maybe just outdoors.

Harington named his invention Ajax and wrote an ode to it. When the Queen visited her naughty cousin months later, she tried the water closet herself—and liked it. Harington supposedly built a similar magical apparatus in the Queen's house, although it's not clear whether it earned him a pardon from his exile or not. In any case, a new toilet system was born, and it was here to stay.[32]

The novel water-powered mechanism took several engineering modifications to become an operational household element rather than a curiosity. Two hundred years later, Harington's design went through several improvements in terms of water management and use. Part of the problem was that even though the toilet removed the waste, it didn't necessarily remove the smell, because the malodorous vapors wafted back into the house through the pipes. Eventually, when a trap containing a water seal to keep out the stink was developed in 1782, the system finally became attractive enough to be widely used. Prudishly named "the water closet" to be acceptable to polite society, the toilet started to gain popularity. British plumber Thomas Crapper began marketing and installing it in the wealthier households. (So if you ever wondered where the words "crapper" and "crap" come from, now you know, and you can use that trivia at your next bar night.)

Having an indoor toilet quickly became a sign of upper-class status. Not only was having a water closet in the house utterly convenient, but it made the uppity lords and ladies appear nearly godlike: they could emerge from their water closets with no evidence left behind, as if proving true the saying that their "shit doesn't stink." They could hide their excrement as if they didn't produce it at all And so indoor plumbing devices became another indicator of class distinction,[33] coveted by all others who couldn't afford them.

The water closet eliminated waste so efficiently and elegantly that an eminent historian of sanitary reform, M. W. Flinn, once said that it "may well have been in the long term the most life-saving invention of all time."[34]

And so, while the Japanese were arguing over who got to keep their valuable poo, the Europeans were happily flushing theirs down the drain. What's more, they were pretending it didn't even exist. Unfortunately, the idea that ignorance is bliss went only so far.

The problem was that while the water closet solved the problem of waste at the household level, it created a massive predicament at a citywide scale. And that massive new problem would soon bite the polite socialites in the very rear ends they were now so comfortably resting on their fancy water closet seats.

CHAPTER 4

THE WATER CLOSET DILEMMA AND THE SEWAGE FARM PARADIGM

In 1884, the Royal Commission on Metropolitan Sewage Discharge, led by a prominent judge, Lord George Bramwell, declared the Thames River a disgrace to civilization. Its dirty banks emitted a stink so potent it overpowered household cesspools. Its banks swelled from the perpetual sewage deposits, interfering with the navigational paths of boats, which ran aground and stuck.[1] Its water was so black that physicist Michael Faraday, who had sailed there in 1854, noted that his white calling card all but vanished from view when barely submerged. An editor at a London publication had apparently tried to use Thames water as ink—and it worked. The ink stunt caused an uproar among the readers and a flurry of outraged letters to the editorial collegiate.

Such unsanitary conditions triggered the infamous epidemics of cholera that swept through England, and the rest of Europe, in 1831, 1837, and 1838. They were eventually traced to water contaminated with sewage. Epidemics of influenza and typhoid followed. But even when the outbreaks abated, the polluted river emitted so many harmful compounds that it still made people sick. In the particularly arid summers of 1858 and 1859, the Thames water level dropped so low that its bared toxic shores began releasing noxious gases—ammonia, methane, and hydrogen sulfide, known for its rotten-egg smell. The scandalous episode was aptly christened as the Great Stink of London. English physician and epidemiologist William Budd, who recognized that some diseases were contagious, described the stink as a historic event. As one newspaper wrote, "For the first time in the history of men, the sewage of nearly three million people had been brought to seethe and ferment under a burning sun, in one vast open cloaca lying in their midst."[2]

One reason the Londoners were so concerned with the smell of the river was that they thought that bad air brought disease. The prevailing medical doctrine espoused the miasma theory, which stated that cholera, typhoid, malaria, and other illnesses were caused by a *miasma*—a noxious, bad-smelling air. In fact, the very word "malaria" comes from the Italian phrase *mal aria*—meaning bad air.

Not surprisingly, Londoners avoided the reeking river like a plague. The boatmen lost customers. Travelers opted for elaborate, miles-long alternative routes to avoid crossing river bridges. The Thames fiasco dominated the press with as much intensity as wars. "India is in revolt and the Thames stinks!"[3] was one prominent writer's pithy description of the nation's problems.

Eventually, neither poor nor rich could tolerate the stink anymore. In 1881, over 13,000 people signed a petition to do something about the problem—and it became the tipping point. Bramwell was appointed to lead a commission to investigate the matter—the very matter suspended in the river's murky waters.[4] It had to identify the nuisance culprit and devise a remedy to the situation.

Tout-à-la-*Thames*

The commissioners bravely undertook an expedition up the river, peering into its dark currents and observing the various objects that floated along. Charged with identifying exactly how far the sewage spread, this aristocratic coterie of knights, lords, and otherwise prominent and respectable officials in tweed suits went on a Thames boat cruise, holding up their pince-nez and pointing fingers at the questionable materials below. In fact, the commission's report said that in the beginning of the journey, the sewage was so thick that its members were actually looking for the glimpses of water in it. "Up to Greenwich it appeared unmixed sewage, then patches of natural water appeared, which increased till the sewage ended," Lord Bramwell said in his report. "We traced the sewage all the way to the Lime-house," he wrote, adding that the offensive matters could also easily reach the London Bridge.[5] He also argued that the public interest required that a "remedy should be applied with the least possible delay."

The culprit behind the problem was the very thing that Flinn called "the most life-saving invention of all time": the water closet.

How did this revered gadget of human innovation turn the Thames River into a disgrace of civilization? The answer is simple: the water closet didn't solve the problem of sewage disposal. It just moved it farther away from people's living quarters. But not far enough.

The water closet had offered the cleanest and most efficient way of removing waste from people's homes. And truth was, it did exactly that. It did flush the excrement out of people's homes and yards. But the problem was that water closets discharged significantly more liquid than chamber pots and privy vaults. As the number of water closets in the city increased dramatically, the small household cesspools, not built to accommodate so much water, naturally began to overflow, especially in heavy rains. The dirty effluent flooded the streets and poured into drinking water sources, spreading cholera and other diseases. The combination of running water, flush toilets, and open sewers proved to be a rather deadly arrangement. It was clear that the real solution lay not only in taking filth out of the house, but out of the city as well. The sewage had to go into some body of water that would take it away—someplace downstream and hopefully far enough that it would no longer endanger humans. The sewage also had to flow through covered pipes, rather than open gutters, that would be buried underground to avoid contamination and prevent the spread of disease.[6]

The Romans had arrived at this wisdom before Christianity had arrived in Rome. The Minoan and Harappan civilizations had built underground pipes that carried sewage away from habitation even earlier. It took humankind a few thousand years to come to the same conclusion again.

The Londoners were not the only ones who arrived at that wisdom. In the nineteenth century, committees of scientists and engineers all over Europe began devising the new type of sewage system that would be robust enough to carry the filth far, far away.

A debate soon ensued, in which scientists and engineers seemed split into two camps. One camp proposed to build two *separate* drain systems—one for rainwater and one for sewage. The first set of pipes would carry the reasonably clean rainwater into the sea, while the second set would transport the filthy sewage someplace else. The other camp favored the idea of a *combined* sewage structure that would rid the city of both, rainwater and sewage, together. The committees even got to discussing where on the garbage map the "kitchen slop" would fit—into the sewage or the rainwater pipes.

Both ideas, it seemed, had merit. The combined sewage system would certainly be cheaper, and could probably use some of the existing structures and pipes. The occasional heavy stormwater torrents would help prevent clogging, making the system "self-cleaning," which was a big concern at the time. Having more water in the system seemed a good thing overall—it would flush things better and probably carry filth farther away. But the separate system camp had valid objections. All that sewage would end up in rivers and seas, they pointed out. Is that what we want? Wouldn't it be more logical to build another system for sewage disposal that would ultimately pump all that humanure to the farms, essentially performing the night men's job in a modern industrialized fashion? It would be so much better to follow the idea of "the rainfall to the river, the sewage to the soil."

The "sewage separatists" were one step away from making waste recycling a reality. If they had, our sewage systems might have been very different today. There was only one problem: no one had any good idea of how to make a humanure distribution system work. The engineers would have to build a system of pipes stretching out from the city to the country, reaching out to all the different farms. Without water as a transport medium, they would have to design a whole new method of pushing poo through the miles-long pipes. That would require some kind of a steam-powered system and energy sources to run it. And after all that work, where would the excrement go at the far countryside end? Would it land in some kind of a cesspool capable of holding the discharge of all London's water closets while the farmers would periodically come over to this "sewer well" with buckets and horses? That may have worked in summer when plants were growing, but what about winter? Obviously, people couldn't stop excreting for a few months, so no matter how big the tank would be, it would eventually overflow. Unfortunately, the dual system idea had too many holes to hold water—or the other thing. It just wasn't technologically possible at the time.

As a result, the other proposal won. The city of London built a combined sewage system, which essentially worked on the French principle of *Tout-à-la-Rue*. Only in this case, the concept had expanded to *Tout-à-la-*Thames. No marvel that the poor river was declared to be a disgrace to civilization. What was surprising is that the civilization around it didn't suffocate and die out altogether.[7]

The sewage system builders had grossly underestimated how much water and sewage would be sent down the pipes. The daily discharge proved beyond immense, far exceeding the original estimates, because Londoners were pulling those convenient flush levers a lot more than the engineers expected. The engineers were also misled by the prevailing scientific doctrine stating that flowing water "purifies itself" and thus would dissipate the filth. By the time the tweed and pince-nez crew took their Thames cruise up to the London Bridge, it was clear that all of the above assumptions had failed miserably.[8]

Unfortunately, neither Lord Bramwell's commission nor the engineers on the project had a good idea on how to resolve the problem. The river could not carry the filth away fast enough. The local farmers didn't want to use the putrid substance even when they were given it for free. Once again, the Europeans didn't see any value in their waste, so they resorted to a very expensive way of disposing of it. Contracted by city authorities, two boats began to carry the sludge out to sea and dump it there—at the exorbitant cost of a million or so pounds. Frankly speaking, at that suffocating point, there were few options left.

Sewage Recycling Resurfaces

After the Thames fiasco, the idea of reusing sewage for farming re-emerged. The commission wrote, "No one doubts, as a general principle, the value of human excreta for fertilising land." It even calculated the "manurial value" of London's sewage, estimating that the discharge of 4 million people would be worth "£1,750,000 per annum." In Japan, this price would have been viewed as giving away fertilizer fortunes for next to nothing, and paying to have it taken away would have been inconceivable.

The irony wasn't lost on Liebig, either. "When we consider the immense value of night-soil as a manure, it is quite astounding that so little attention is paid to preserve it," he lamented. "The quantity is immense which is carried down by the drains in London to the River Thames, serving no other purpose than to pollute its waters."[9]

Even the great writers of the time bemoaned the foolish wasting of our waste. "What do we do with all this golden dung?" questioned Victor Hugo, describing the Parisian cloaca in his classic *Les Misérables*. "It is swept into an abyss. All the human and animal manure which the world wastes, if returned to the land, instead of thrown into the sea, would suffice to nourish the world."

The prominent sanitary thinkers of the time looked at how other countries and societies waged their excrement battles. Some pointed to the New World, where, in contrast to the great Thames fiasco, excrement recycling was actually enjoying a moment of success.

In the 1830s, New York City paid New Jersey–based Lodi Manufacturing Company to rid the growing metropolis of all the excrement it produced. The process proved to be somewhat similar to those of Japan or Flanders. Even the name "night soil" remained the same, although it was supposedly called that not because the chamber pots were put out the door in the morning, but because the scavengers collected it at night, to spare residents from the stink.

The scavengers gathered the muck in barrels and dumped it into Lodi's boats. Lodi dispatched the excrement to its processing plant, where it was dried, deodorized, and turned into a fine powder charmingly called *poudrette*—French for fine dust. That dust was then mixed with other organic substances—including Peruvian guano, or bird shit, which also made good fertilizer—and then shaped into briquettes and shipped to farmers. The product, which Lodi named *tafeu*—which supposedly meant "processed night soil" in Chinese—was in demand.[10]

Lodi may have owed this recycling success to the fact that, compared with London, New York was tiny at the time—its population fluctuated around 200,000 people. Another reason was, once again, the condition of the soil. Farmers in nearby Long Island had to make do with sandy soil, which wasn't very favorable for growing food. So they used every scrap of organic matter they could get—manure, excrement, and ground bones—to boost their yields, then sold their produce to the Manhattan and Brooklyn residents who had helped fertilize it, much like Japanese farmers. Other farmland around large urban areas was also becoming depleted, so fertilizer was in demand.[11]

Tafeu made good business because Lodi essentially made money twice. First, the city paid it to remove the muck, and then it sold its *tafeu* to farmers for 40 cents per bushel. In fact, business was so good that more companies joined the *poudrette* and *tafeu* business, and similar recycling methods sprang up in Philadelphia, Boston, and Baltimore. American agrarian publications—*Country Gentleman*, *Farmer's Cabinet*, and *American Agriculturist*—advocated the use of humanure and insisted that every city should produce it. Farmers and agriculturalists debated the most efficient and potent manure preparations. Some thought that the *poudrette* method depleted the valuable fertilizing essentials by as much as half. Fresh unprocessed shit, they argued, was best.[12]

The inventive Yankees didn't stop there. They wanted to automate the city-to-farm sewage delivery process. The idea would kill two birds with one stone: it would continuously pump fresh shit from cities to fields, and it would take the human out of humanure handling altogether. The Americans tried building a sewage-fed farm—and succeeded.

The Bright Sewage Farm Future

In the late nineteenth century, American railroad engineer George Mortimer Pullman, who made a fortune on his Pullman sleeping cars, financed a rather unusual project for his time. He decided to build a novel state-of-the-art sewage recycling system for Pullman, his company town, which he founded on Lake Calumet, not far from Chicago. Today, Pullman is a part of Chicago, but back then it was a small manufacturing town of about 8,000 residents, surrounded by farmland. Pullman concluded that the shallow lake, with its weak currents, wouldn't be able to absorb the town's sewage, so unlike the London engineers, he resolved to build two separate systems. One would drain rainwater into the lake, and the other would channel sewage outside the city—and onto the fields. This idea, widely discussed at the time, was called sewage farming.

The concept of sewage farms had percolated in the minds of sanitary thinkers for a while—both in Europe and in the New World. The idea was that sewage effluent would be pumped from a city to a nearby farm. In one version of the concept, the effluent would be sprayed directly onto the land via a sprinkler—think artificial sewage rain. (This may sound horrific, but would it really be much worse than the toxic pesticides we spray on the fields today?) In another version, the flow would be distributed through a system of canals and ridges directly into the soil, where plant roots would soak up the nutrients. Either way, the nutrients would be returned to the earth, the plants would use them to grow, and the water would percolate through the soil or evaporate, becoming reasonably pure and suitable for drinking again. In any case, it would have been much cleaner than water fetched from the Thames, or any other big-city river of the time.

Built by engineer Benezette Williams, Pullman's sewage farm project went into production in 1881. Propelled by a steam engine, the waste of the town's residents traveled through a three-mile-long pipe, and was then used to fertilize a farm. Within a decade, the system was pumping almost 2 million gallons of sewage daily, irrigating about 140 acres of land. Allegedly, by the end of its journey the substance didn't emit particularly noxious odors—which, once again, was a very important aspect of hygiene, according to the miasma theory.[13]

The idea of putting modern sewage sludge directly onto farm fields, with all the chemicals floating around it in, sounds terrifying.

But the truth is, back then, the sludge pumped through Pullman's pipe was rather superior to ours—at least in the agricultural sense. It was free from chlorine and other harmful household pollutants, because the chemical industry hadn't created them yet. It had no antibiotics, because Alexander Fleming had yet to discover one. It also contained no Prozac, no BPA-laden plastics, no phthalates, and none of the other chemicals and pharmaceuticals we created in the century that followed. There may have been some heavy metals, parasites, and bacteria in it, but let's face it, we haven't exactly cleaned up our own food supply today. By nineteenth-century standards, Pullman's method was closer to sustainable, renewable agriculture than most of our planet is today. At about the same time, a few other sewage farms sprouted roots across America—in Texas, Nebraska, Rhode Island, and Southern California.

Inspired by Pullman's example and a few others, the prominent British lawyer and social reformer Edwin Chadwick wanted to build a similar system in London. Having devoted his life to sanitary improvements, Chadwick was a fierce advocate of building separate sewage systems that would feed into sewage farms. He believed that when designed right, these farms could yield produce of extraordinary quality in massive amounts. Chadwick reported that Pullman's farm yielded 200,000 heads of cabbage, 18,000 bunches of celery, 100 tons of hay, and a great deal of other farm produce. He also wrote that he was "informed that an important example of a separate system is in progress for Baltimore."[14]

Chadwick's challenge was that London was nearly three orders of magnitude larger than Pullman. The little lakeside town had 8,000 inhabitants. London, on the brink of the nineteenth century, had over 6 million. A London sewage farm system would have to operate on a much grander scale—one three-mile-long pipe wouldn't do. So Chadwick proposed a radiating system of pipes that would carry the sewage out of the city to sumps—pits that would accumulate it, and from which it "would be ejected, by engine power, by steam power" to be used as liquid manure. According to his estimates, an acre of land would be able to absorb the sewage of about a hundred people—although he also speculated that even less land would suffice. Chadwick envisioned a great many uses for that pumped fertilizer supply. "Besides farms of an extraordinary productive power, the fresh liquefied manure principle of culture would serve for the institution of parks and pleasure grounds in advance of any that horticulture has yet attained." He even sketched out the pricing of that piping system. "The estimated cost of the culverts and channels for carrying it that distance was twelve hundred thousand pounds, or fifty thousand pounds a mile."

The idea of sewage farms was so alluring that some British royals rolled up their sleeves and dug into dirt to prove the concept. Chadwick described horticulturist Sir Joseph Paxton as a successful "liquefied manure farmer." "In such hands, by horticultural feeding, the carrot has been made a new plant, with finer saccharine matter and a new aroma. The celery is most excellent. The rhubarb is made to exceed itself; it has been used for champagne." Even the animals seemed to favor produce grown with humanure. "A cow was selected, and sewaged and un-sewaged grass was placed before it for its choice," Chadwick wrote. "It preferred the sewaged grass with avidity, and it yielded its final judgment in superior milk and butter of increased quantity."[15] Whether the latter was true or not, Chadwick was prompt to point out the ridiculousness of dumping all that waste into the river and causing the Thames fiasco. "It is with such raw material of production that the superior Legislature has allowed the vestral local authorities to pollute the river."[16]

The rest of Europe, too, jumped on the sewage farm bandwagon. In some places, and for certain periods of time, these sewage farms actually did work—in Edinburgh, Paris, Danzig, Milan, and Croydon, a town in England. The most successful one was in Berlin. The farms grew all kinds of produce, from cabbages to strawberries, whose size and quality were described as remarkable.

Some particularly successful farms hosted gala luncheons cooked entirely from their sewage-fertilized crops. Besides the vegetables, the menu included beef from cattle raised on sewage-fertilized grass and cheese made from their milk. It also included the final, purified effluent, served in crystal decanters. That last bit may sound insane, but apparently when typhoid swept through Berlin, none of its sewage farm workers fell ill, despite drinking that very effluent with their lunches.

Turns out, this outcome isn't really surprising at all. On the contrary, it's rooted in hard science—In the complex physics and chemistry of soil—even though that complexity wasn't understood at the time. "That's the beauty of soil—its ability to naturally purify water," explains soil scientist Mary Stromberger. "It's one of the most important ecosystem services of soil."

As water trickles through, soil purifies it both chemically and physically. Yes, the soil contains organisms pathogenic to humans, but it also harbors huge numbers of beneficial bacteria and fungi. These species have evolved to degrade all kinds of compounds harmful to us, but useful to them. They may be using these compounds as a carbon source for energy or other needs. Some of these species can even break down pesticides and chemicals from agricultural runoff. They naturally biodegrade these compounds.

Organic matter is also sticky. So as water percolates through soil, the various compounds present in it stick to soil particles. The result is a clean effluent. It's like the charcoal filters used in aquarium tanks—charcoal is just another activated form of organic matter. "Soil functions as a giant filter that does chemical and physical filtration," Stromberger says. "And that's the premise for using soils for purifying sewage waste in septic systems—a classic example of how soils have been used to purify polluted water. You see that a lot in rural environments."

That's why, during the typhoid outbreaks, the sewage farm workers may have been better off than their urban counterparts. Their soils filtered water better than the urban soils, which were probably overloaded by sewage and other pollutants.

Unfortunately, not all sewage farms worked well. Chadwick noted that farming with liquefied manure was a delicate thing. In order for it to work, the farmers had to know how much of it to apply, and how often. When untrained village folk poured too much of it onto the plants, it would cut off the oxygen needed by the aerobic soil bacteria to decompose waste. As a result, the land would become "sewage sick." Different, anaerobic, bacteria would proliferate, which under these conditions wouldn't break down waste as efficiently, and would produce offensive smells. Running sewage farms required great expertise, which the Japanese may have perfected, but few European farmers possessed.[17]

Because there was a limit to how much manure farmland could absorb, some sewage farms indeed became saturated and collapsed. The big metropolises produced so much sewage that there simply wasn't enough arable land around to absorb all that shit. As Christopher Hamlin, a historian of science and technology at Notre Dame University in Indiana, wrote, "Under European conditions, treating the sewage of 100,000 people required at least a square mile of land. . . . Finding the necessary 30 or so square miles in the vicinity of a city like London, for example, was out of the question."

Part of the Pullman farm's success was due to the fact that it operated on a small scale. It ran for over two decades, and was still profitable in 1894. Shortly after that, Pullman's manufacturing empire hit hard economic times. George Pullman died in 1897, and the farm eventually stopped functioning. Other sewage farms withered, too. Despite being the closest thing to the ideal waste recycling process, they eventually fell out of favor. It was easier to dump waste into the water. Plus, humans found other ways to enrich soil. For some time, people used guano—nutrient-rich bird droppings scraped off remote oceanic islands where they had accumulated over centuries—which boats brought to Europe and the United States. But as humankind crossed over into the twentieth century, two important new developments took hold: the germ theory and the advent of commercial fertilizer. And that changed everything.

CHAPTER 5

GERMS, FERTILIZER, AND THE POOP POLICE

Two hundred years before the Great Stink, the Royal Society of London received a peculiar letter from a Dutchman named Antoni van Leeuwenhoek. A businessman turned scientist, Leeuwenhoek claimed that he had discovered a great number of minuscule creatures in a mere drop of water. Moreover, he cited several credible people who had confirmed his discovery, and outlined the instructions for replicating his experiments. The secretary of the Royal Society, Robert Hooke, was intrigued.

It all started in April 1676, when Leeuwenhoek was playing around with a magnifying glass. He had previously worked in a drape shop and had learned to make his own lenses to see the threads of fabrics better. That got him into the business of microscopy. Plus, Leeuwenhoek was very curious. He might have been one of those people whose curiosity doesn't wane after growing up. So he was interested in just about everything. At the moment, he was into the intricacies of taste. Having been "sickly and nearly unable to taste," he was wondering where that gustatory sense came from, and why certain spices, including ginger, nutmeg, and pepper, retained their taste for months.

In one of his explorations, Leeuwenhoek placed some pepper in water for some time, and eventually stuck the mixture under his glass. But he saw a whole lot more than the original spice. He saw tiny organisms energetically swimming around.

"This pepper having lain about 3 weeks in the water, to which I had twice added some Snowwater, the other water being in part exhaled; I looked upon it the 24 of April 1676," he wrote, "and discern'd in it, to my great wonder, an incredible number of very little animals of divers kinds." To Leeuwenhoek, the creatures must have looked a bit like eels, because he named them "little eels," or *animalcules*.[1]

He immediately turned his glass on other substances—such as a piece of white matter stuck in his teeth. "I have mixed it with clean rain water, in which there were no 'animalcules,'" he wrote, "and I almost always saw with great wonder that there were many very little *animalcules*, very prettily amoving."[2]

To ensure that his findings were correct, Leeuwenhoek recruited others to observe the animalcules, and eventually sent off his letter to Hooke. "This exceeds belief," he wrote, but "I have here sent the Testimonials of eight credible persons; some of which affirm they have seen 10000, others 30000, others 45000 little living creatures, in a quantity of water as big as a grain of Millet."

Hooke, whose job at the Royal Society was to verify experiments, attempted to replicate Leeuwenhoek's findings, but didn't see anything at first. "I concluded therefore that either that my Microscope was not so good as that he made use of, or that the time of the year (which was in November) was not so fit for such generations, or else that there might be somewhat ascribed to the difference of places; as that Holland might be more proper for the production of such little creatures than England." A few days later, Hooke looked again—and lo and behold, the creatures were there. "I examined again the said water; and then much to wonder I discovered vast multitudes of those exceeding small creatures, which Mr. Leeuwenhoeck had described."[3]

Leeuwenhoek's magnifying glass experiments helped launch not one, but several scientific disciplines—microbiology, bacteriology, immunology. He gave humankind the ability to discover all kinds of microscopic organisms living in all kinds of places—in water, on plants, and on humans.

Leeuwenhoek's findings spearheaded an ardent interest in all microorganisms, and so more scholars began to peer through lenses in zealous efforts to identify more mini-creatures and learn what they do. As scientists built their understanding of health and disease, they began to realize that many microorganisms possessed immense powers.

In the mid-nineteenth century, French scientist Louis Pasteur, who developed early vaccines and taught the world that pasteurizing milk kills bacteria, performed many experiments that showed why wine and dairy products go bad. Pasteur proved that bacteria were to blame. He also proved that the little buggers could not only spoil our food, but also ruin our health. Other scientists also realized that some of the cute "animalcules, very prettily amoving" in a drop of water, or elsewhere, could be deadly.

Throughout the second half of the nineteenth century, just as Londoners, Parisians, Bostonians, and other urban dwellers were fighting diseases like cholera and typhoid spread by water contaminated with feces, the science of microbiology was rapidly maturing. The world of bacterial pathogens was unfolding with frightening speed. In 1854, Italian scientist Filippo Pacini discovered *Vibrio cholerae* (although he didn't get recognition for it until 82 years after his death, in 1965). Three years later, Theodor Escherich, a German-Austrian pediatrician, zeroed in on *Escherichia coli*. At about the same time, Theobald Smith, an assistant to American veterinary surgeon Daniel Elmer Salmon, identified *Salmonella choleraesuis*, naming it in Salmon's honor.

Many of these bugs were commonly found in feces, manure, and dirt. So as the germ theory of disease replaced the miasma theory, excrement became vilified more than ever. Scientists, chemists, agriculturists, and farmers began to reconsider their views. Who in their right mind would put this dangerous, hazardous, deadly substance onto their fields? And why would anyone eat the produce grown on a farm that used humanure? Hadn't humankind advanced enough to figure out other, safer ways of fertilizing its fields?

The importation of guano also added to the devaluation of human excrement. And even though the guano deposits on oceanic islands would eventually be depleted, humans continued to look for fertilizers other than their own organic power. The evil microbes and the rising powers of chemical science made it obsolete. And just like that, our excrement's reputation as a resource was doomed. Humans no longer needed the fertilizer they so easily produced themselves.

Better than that, they created a way to make fertilizer quite literally out of thin air.

Enter the Haber-Bosch Process

All plants need nitrogen to grow and build their cell walls. Liebig, the German chemist, came to this conclusion in the 1840s. But nitrogen is tricky. While there may be plenty of it stored in decaying organic matter within the soil, its availability to growing plants depends on how quickly recycling microorganisms in the soil decompose that organic matter. Furthermore, the rhizobia—those ingenious bacteria that can grab nitrogen from the atmosphere and use their nitrogenase enzyme to convert it into plant food—are fussy. They can do that work only within the roots of legume plants, where they are protected from oxygen exposure. The rhizobia form symbiotic relationships with the plants, providing both the plants and the soil with nitrogen. And while rhizobia do this conversion so expertly, humans couldn't master any comparable process until the early twentieth century.

Nitrogen is so abundant that it constitutes about 78 percent of the air all around us. But it is a highly inert gas. A nitrogen gas molecule (N_2), composed of two nitrogen atoms, is so stable that for a long time chemists couldn't devise a way to grasp it from its gaseous state, break it apart, fix it in a more manageable form, and then apply it to the soil. Because of its remarkable stability, atmospheric nitrogen can react with other elements only in the presence of high pressure, high temperature, and special catalysts. In other words, this reaction requires a very powerful technology. It wasn't something that Liebig or his contemporaries could do. But about 60 years later, two scientists in Germany, Fritz Haber and Carl Bosch, managed to create such technology—and it went on revolutionize the world.

Working in his lab, Haber figured out the right combination of temperature, pressure, and catalysts needed to get atmospheric nitrogen to react with another gas. He circulated nitrogen and hydrogen over a catalyst at 500°C and under a pressure of 150–

200 atmospheres. (By contrast, the pressure at sea level is measured at 1 atmosphere.) The reaction of the two gases synthesized ammonia—NH_3—which makes great fertilizer. The reaction was so volatile that no man-made chamber could contain it, so Haber initially achieved it within a hollow quartz crystal, says Mark Benvenuto, professor of chemistry at the University of Detroit Mercy, who studied the history of this process. That container worked for a little tabletop apparatus, but no quartz crystal was big enough to accommodate an industrial-sized operation.

German chemical company BASF purchased the rights to the process and paired up Haber with its engineer Carl Bosch to scale up and commercialize his technology. Achieving that goal proved tricky: "All iron chambers would blow up after three days," Benvenuto says. Eventually, the duo figured out that the chemical reaction was making the iron brittle and came up with a clever solution that involved nested chambers. When commercial production of fertilizer began in the second decade of the twentieth century, the method went down in history as the Haber-Bosch process.

The Haber-Bosch process was resource-intensive and polluting, and although Haber won a 1918 Nobel Prize for its discovery, it also led to some very dark moments in human history and caused humankind much harm. Haber used the same nitrogen-fixing technology to create explosives during World War I. He also created the world's first chemical weapon: the chlorine gas Germany used on its enemies during the war.[4] More terrifyingly, his work on agricultural insecticides led to the creation of the infamous poisonous gas Zyklon B, which the Nazi Reich employed in its concentration camps all over Europe during World War II, killing millions. Haber didn't live to see that application of his method. He died in 1934 after a long illness, triggered by the fact that as a Jew in Germany, he was no longer allowed to continue his scientific work.[5]

But the Haber-Bosch process lived on. And it gave humankind an unprecedented ability to grow food. Along with vaccination, antibiotics, and other medical advances, it has most certainly contributed to the world's population explosion, in which humankind grew from a little over a half billion people in 1900 to 7 billion today. Vaclav Smil, professor of geography at the University of Manitoba and author of the book *Enriching the Earth*, which traces the history of fertilizer, writes that "without that synthesis, two thirds of the world's population wouldn't be around."[6] They simply wouldn't have enough to eat. In fact, they wouldn't have been born, because their parents wouldn't have had enough to eat, either. According to Smil's book, humankind synthesizes about 130 million metric tons of ammonia a year, and 80 percent of it goes into creating fertilizers.

With the remarkable success of synthetic fertilizers, humanure wasn't needed at all. Why would anyone deal with that disgusting substance when one could simply buy a load of decent-smelling, pathogen-free, synthetic goodness? Furthermore, synthetic fertilizers also seemed to work better—they generated bigger yields with less toil. The cultural shift to synthetic fertilizers took hold not only in the Western world, but also in China. As Worster described in "The Good Muck," "The old manure-based farming was hard work for an old man—including so much painful stooping to insert rice plants into the paddies, then the meticulous spooning out of liquid and biosolid manures, and finally the harvesting and threshing of the rice crops." Tired of that physical labor, the old man tried synthetic fertilizer—and that proved so easy, cheap, and effective that he was soon tossing around about 500 pounds of nitrogen a year, and "his yields more than doubled to 7,200 pounds (3,266 kilograms) per acre." Today, China is one of the largest fertilizer consumers in the world. Some 80 percent of the nitrogen in Chinese bodies now comes from food produced with the aid of chemical fertilizers.[7]

And so, in the early twentieth century, humanure dropped precipitously in value. A substance treasured like gold by some cultures and seen as important, albeit unpleasant, matter by others was vilified and downgraded to the status of unmentionable refuse by industrialized societies. We no longer needed it. And worse, we feared it. We were afraid that it was out to kill us or make us sick. And so, from that point on, all our excremental battles would be directed at removing, destroying, or annihilating it as quickly and efficiently as possible. The natural organic substance we all produce daily was officially transformed from a commodity to human waste. Our modernized society had no use for it whatsoever.

The American Poop Police

Just as the upper-class Victorian Londoners felt godlike with their water closets, so did the progressive Americans on the brink of the twentieth century. Only they went one big step further in their excrement wars: they were building the great American sewage systems, and they made sure everyone was on board, whether they liked it or not.

In his PhD thesis, aptly titled "American Wasteland," Daniel Max Gerling, director of the Writing Center at Augustana University, explains that for American society, this modernizing experience was twofold. On one hand, we domesticated our excrement—we brought it from outhouses and latrines into our homes. On the other hand, enlightened with the knowledge of all its evils, we became obsessed with expelling it from our dwellings as quickly as we could. "The odd paradox of this period," Gerling writes, "is that by entering the home, excrement was met with far more radical efforts to expel it forever." Having clean toilets that eliminated poop instantly was a sign of sophistication. And because, traditionally, it was women who kept the house clean, toilet maintenance naturally fell under their domain.[8]

Americans spread their way of thinking to other cultures and places that had no organized waste management—for example, to Native American reservations. Installing sewage systems was not a straightforward task, so instead the residents were taught to build latrines. Women played a significant role here, too. In 1890, the United States Bureau of Indian Affairs (BIA) hired thousands of young women to travel to the reservations, visit people's homes, and instruct the residents how to keep their houses and outhouses clean. These matrons dropped in and explained why squatting underneath a tree was wrong and squatting over a latrine hole was better. And they diligently recorded latrine conditions in their field questionnaires. Overall cleanliness earned 10 points, but clean excrement disposal was worth 50. With so many strict standards to enforce, they were essentially poop police, Gerling says.[9]

The people who were told to change their toilet habits didn't like being policed and often resisted. Gerling writes that even children who attended boarding schools, where learning the proper toilet etiquette was part of the program, protested—by means such as relieving themselves on the floor. Other schools were more successful. Gerling quotes one student as saying, "I was uneasy at first and expected the bowl to overflow; but I caught on quickly and like it." He also added presciently, with a far-reaching wisdom that we can certainly appreciate today: "Although it was a waste of water."[10]

The High Price of Heading for the Bush

This sanitary policing was unpleasant, but building sewage systems on a nationwide scale is what shielded the Western world from the epidemics of diseases and endemic parasitic infections caused by raw sewage. In other parts of the planet where centralized sanitation systems have not been built—whether due to lack of resources, water scarcity, or other obstacles—these diseases continue to take a huge toll on human health. The World Health Organization (WHO) estimates that 2.4 billion people on this planet still lack access to basic toilet facilities, and that nearly 1 billion still head for the bush—to fields, street gutters, or creeks—what epidemiologists euphemistically call "practicing open defecation." When that fecal matter leaches into the drinking water supply—quite often traveling as far as the next village downstream—it spreads cholera, dysentery, intestinal worms, parasitic infections, and even polio, fueling the never-ending cycle of disease and poverty.

According to WHO's 2016 numbers, about 842,000 people in low- and middle-income countries die every year due to poor water quality, sanitation, and hygiene. The American Centers for Disease Control and Prevention (CDC) estimates are even higher. The CDC statistics say that worldwide, over 2,000 children die from diarrheal disease daily[11]—more than from HIV/AIDS, malaria, and measles combined. Among the first preventative measures listed on the CDC website are providing safe water and providing adequate sanitation and human waste disposal. Epidemiologists say that depending on the country, if you add up the health care costs, missed work time, school absences, and other lost opportunities, the economic toll of inadequate sanitation can vary from 1 to 4 percent of the country's gross domestic product, and in some cases is even higher. Young women drop out of school because they have no place to pee safely, let alone to change their pads or cloths when they menstruate.

Not every society can easily implement a functioning sanitation system, however. In many places, building Western-type flushing toilet systems that rely on a steady supply of water is not possible. Some places simply don't have enough of that convenient flushing medium. Others have too much, which isn't good either. Too much water creates floods, overflows, and sewage spills. Some countries don't have the financial resources to invest in a massive countrywide system upgrade. Others are struggling with aging or failing systems that can no longer support their burgeoning cities. Even when basic toilets like latrines are built, they don't necessarily make for a functioning, hygienic solution. They break, leak, and must be regularly emptied, which is usually done manually and is far from a hygienic operation. In some places it's done by men with buckets, who often drink heavily to numb their senses or pour kerosene onto the waste before descending into the pits.

The variety of pathogens that can thrive in latrine sludge is beyond imaginable. Name a dangerous germ, and it's probably hanging out in the latrine pit, or around it: hepatitis A and B, campylobacter, cholera, dysentery, salmonella, *E. coli*, and every intestinal parasite known to science. Not to mention the intestinal worms that you don't even have to swallow to get sick because some of them can become airborne—inhaling them is enough to get you infected. Some can quite literally worm their way in through one's bare feet. Professor Francis de los Reyes at North Carolina State University, who has taken samples from latrines all over the world for his environmental and epidemiological research, has seen all of that and then some. He routinely takes a deworming pill when he gathers samples—in addition to being vaccinated. But most people who empty latrines don't take deworming pills. Some of them don't even wear gloves.

In India, women of the lowest caste, the Dalits, empty the latrine pits. In India's outlawed but still persistent caste system, Dalits are considered the lowest of the low by the higher classes. Their job is to clean the so-called dry latrines of the rich. As the name suggests, dry latrines have no water flowing through them and no pits underneath. They are small compartments where the rich relieve themselves, leaving the filth on the floor or the ground. The Dalit women come with buckets and scoops, gather up the waste, and take it away, carrying it in woven buckets on their heads—a mind-boggling picture painted by journalist Rose George in *The Big Necessity*.[12] They do this work in the blistering sun and the pouring monsoon rains, and they are paid a pittance for their labor—sometimes barely more than a serving of bread. They are shunned and often abused and beaten when they try to fetch water from communal wells or buy food at the market. Yet they pass their profession on to their daughters, teaching them the job from a young age—and perpetuating the outcast cycle they can't break as long as the archaic toilets remain in existence.

Indian sociologist Bindeshwar Pathak believed that populating India with toilets that didn't require humans to clean them would free Dalits from their centuries-old abuse. So he designed the Sulabh, an ecological composting toilet that uses only a liter and a half of water to flush and has two alternating pits. When one pit fills up, it's closed and left to compost its content, while waste accumulates inside the other one. The name Sulabh means "easy to get," and indeed, over 1.5 million of these toilets have been installed in India. The project has freed hundreds of thousands of Dalits of their horrific obligations, while also offering them retraining for other professions. The women's stories—each a unique personal narrative, but with a common theme of finally ditching the stigma—have been told in the book *New Princesses of Alwar: Shame to Pride*, which I picked up at the Sulabh International Museum of Toilets in New Delhi.[13] "These women had such a transformation!" museum curator Manoj Kumar told me.

De los Reyes, who designs sanitation management solutions for developing counties (read "machines that empty latrines"), says that automating the process of latrine emptying can solve many of these problems. But the automation doesn't solve humankind's most basic and oldest waste management problem: What do we do with all this shit? No matter how far away we take it, it causes disease. No matter what body of water we put it in, it will come back to us. With the 7 billion people populating this planet right now, it's just not possible to dump our waste someplace and ignore it, says Canadian epidemiologist David Waltner-Toews. "In the 21st century, everyone is downstream from everyone else."

We may think that we have solved the excrement problem in the Western world with our massive sewage plants. But the bitter truth is that we have solved only one problem—ensuring that our excrement no longer endangers our health. We no longer die in large numbers in epidemics of waterborne disease, and we have essentially eliminated parasites like intestinal worms. But by solving the health problem, we've created another one—a huge environmental predicament, which has taken decades to develop.

Unlike the cholera epidemics that swept through nineteenth-century cities like wildfire, this new problem is more like a slowly ticking time bomb that took over a century to register. Now, two decades into the new millennium, we are finally starting to hear its measured and unforgiving tick. And unless we do something really soon, that bomb will blow up, leaving our already fragile ecology in a possibly irreparable state.

PART 2
THE PRESENT: A SLUDGE REVOLUTION IN PROGRESS

CHAPTER 6

THE GREAT SEWAGE TIME BOMB AND THE REDISTRIBUTION OF NUTRIENTS ON THE PLANET

"Lina and Craig, you take the net, go into the water, and catch some crabs and fish," calls Linda Deegan as she dispatches us to do some scientific fishing. Craig LeMoult, a fellow journalist from WGBH News, and I pick up the net and head down to the water.

Our other colleagues are put to work, too. Sarah Kaplan, from the *Washington Post*, and Denise Hruby, from the Shanghai-based English language newspaper *Sixth Tone*, get to pluck mussels from the pond's muddy banks. The last reporter duo, freelancer Marcus Woo and María Mónica Monsalve, from Colombia's *El Espectador*, are sent to draw groundwater samples uphill.

This is not the type of work journalists normally do. Usually Deegan, an aquatic ecosystems researcher at the Marine Biological Laboratory in Woods Hole, Massachusetts, would be taking her postdocs and graduate students on these quests. Today, she is accompanied by the six lucky journalist fellows selected by the MBL to dig in the mud right next to our teachers for a week to learn what it means to be a scientist. We are much less knowledgeable and helpful than Deegan's average students, plus we have questions, often completely random and off topic. Working with us is like herding extremely curious cats. But we try to be productive. It's not every day that writers get to investigate one of the major slow-brewing calamities wrought by humanity—the perilous effects of our sewage on the planet.

We are measuring nitrogen accumulation, or *"nitrogen loading,"* in soil, water, and living creatures around various areas of Cape Cod.

Known for its quaint villages, bays, ponds, and beaches, Cape Cod is a favorite New England summertime destination. But Cape Cod has a dark side, too—nitrogen leaches from the septic systems of its over-built summer housing developments. This nitrogen pollution is essentially a slowly ticking ecological time bomb that most vacationers here have never heard about. We six journalists want to understand what led to this situation and how and why it happened.

Septic systems are designed to accumulate solid sewage and let the liquid seep into the soil, from which it slowly dissipates through groundwater. That effluent contains a lot of urine, which is high in nutrients like phosphorus and, especially, nitrogen—that miracle fertilizer Haber and Bosch worked so hard to synthesize. When that nutrient-rich effluent seeps from Cape Cod's septic tanks into streams, rivers, and swamplands, it overfertilizes every body of water it reaches, throwing it out of balance. It is that potent efflu-ent that causes algal blooms, plant overgrowth, coastal marsh decay, and other ills. In scientific terms, this problem is called "*eutrophica-tion*," which basically means "nitrogen overload."

Deegan's work on Cape Cod centers on understanding how this man-made nutrient surplus affects the coastal marshes, and we are here to help her gather and compare information on nitrogen load-ing at three different Cape Cod sites: Sage Lot, the head of Waquoit Bay, and Quashnet Pond. Spared from development, Sage Lot is a pristine, intact coastal marsh. Waquoit Bay, which boasts some resi-dential homes with septic tanks, is somewhat affected, but is still holding up. The Quashnet site, which has soaked up a lot of nitrogen, is a mess. Part of it is a swampy, decaying, grassy mud flat, and part is indeed a pond, albeit with collapsing banks that are so slippery you must be careful not to fall into the foul-smelling water, which looks dark blue and opaque like ink. Quashnet Pond is where Craig and I have been told to cast our seines to catch marine creatures for Deegan's nitrogen measurements.

With waders and gloves on, I try to inch slowly into the water from the bank. Instead, I skid into the pond with a splash, and my feet im-mediately sink into the soft, muddy bottom—the physical experience of a coastal ecosystem in crisis. "Careful," warns Rich McHorney,

Deegan's colleague and a jack-of-all-trades who, with equal ease, can give you a lecture on Cape Cod's glacial geological history and fix a malfunctioning CO_2 gauge box that looks like a 1970s relic. "Try doing it slow," Craig heeds his advice, but he, too, slides down with a splash. We regain our footing and stretch out our seine.

After a half hour of fishing, we climb back onto the muddy banks and examine the catch. In the net are fiddler crabs, tiny silvery fish called mummichogs, and piles of sea grass. As we sort the specimens into different containers and plastic bags, a huge crab crawls out of the pile and snaps at Craig's finger. "Dinner!" jokes the rest of our group. We will dissect the catch and send it to a lab to assess the amount of nitrogen in it.

Dinner, in fact, is still a ways off. So far, we've drawn up groundwater, dipped into creeks, and waded into the sea. "Now we will do something really truly amazing," McHorney says. "We will listen to how marshes breathe."

Before we can start asking questions, he hands us some strange-looking equipment consisting of plastic rings and lids with gauges that fit on top of the rings. He also gives us instructions: "Place these rings on the ground, cut the grass within the circle, and pluck out any marsh creatures crawling in it. Then cover the ring with a lid and measure how much carbon dioxide the marsh is exhaling."

The last tidbit isn't a joke. Marshes, as well as meadows, forests, and other ecosystems, do indeed breathe. A healthy forest removes carbon dioxide from the air, replacing it with oxygen. So does a healthy coastal marsh. But decaying, rotting forests or marshes respire differently. They produce a bunch of carbon dioxide because they rot—and so they contribute to the infamous greenhouse effect. The more nitrogen is loaded into the soil, the more carbon dioxide the marsh exhales.

I am captivated by the thought of listening to the marsh breathing. You can't really "hear" it, but once you cover the circles with the gauge lids, you can measure the amount of carbon dioxide that accumulates in the little round spot underneath over time. And then scientists can compare how well healthy marshes and decaying marshes sequester carbon. Measuring the marsh's exhalations keeps

FIGURE 2. Waders on and pens out, journalist fellows at the Marine Biological Laboratory are taking notes from Rich McHorney on setting up equipment to listen to marshes "breathing"—recording how much carbon dioxide they release into the atmosphere. CREDIT: LINA ZELDOVICH

us crawling in mud on all fours for a couple of hours. Deegan has been crawling in mud for over a decade, so she knows what it's like. She was the one who set the stage for understanding what our sewage does to our waters and lands.

How the TIDE Project Discovered the Great Sewage Time Bomb

Today, ecologists know that nitrogen leaching from our sewage systems and our agricultural fields is the culprit behind algal blooms and coastal marsh decay. It was Deegan who figured it out. In the spring of 2003, she and a dozen other energetic ecologists and graduate students disembarked from a research boat onto Plum Island, off the northeastern coast of Massachusetts. The scientists brought with them a huge amount of nitrogen fertilizer—loads of 50-pound bags, which they hauled from the boat to the island's salt marsh. When the bags were finally stacked into a neat pile, the scientists snapped a celebratory group picture, took a breather, and set to work. They mixed the fertilizer into a concentrated solution, then used a computer-controlled system to pump the potent cocktail into the water that naturally flooded the marsh with every tide. They proceeded to do this all summer long, schlepping in 40 bags every three days. "At the beginning of the season, we celebrated the first mix with Mimosas on the Marsh," Deegan says. Mimosas became a tradition, and bag schlepping a summer workout. "At the end of the season, we were tired, but strong."

"In 2003, the scientific community viewed salt marshes as nitrogen sponges that could absorb nearly endless amounts of it from the water," Deegan says—because the marsh plants would simply use it to grow, and what harm could that do? "The dominant paradigm was that excess nutrients can't hurt marshes and might even help. So the bottom line was that we didn't have to worry about releasing our nitrogen-rich sewage effluent into the sea."

Deegan was skeptical of that view and wanted to test it out. But the doctrine was so well established that she had trouble getting funding for her proposed TIDE Project (TIDE stood for *T*rophic Cascade and *I*nteracting Control Processes in a *D*etritus-Based Aquatic *E*cosystem). Her proposal got turned down twice, but she was persistent, and in 2003 she began her work.

"And so we set off to investigate," Deegan says. The team turned a six-hectare parcel of land into a unique outdoor lab and kept other land as a "reference marsh" that wasn't nitrogen-fed. According to the established view, nitrogen overload should've made marsh grasses grow out of control, turning the swampy land into a lush meadow.

But that's not what happened.

Within a couple of years, the team members noticed some peculiar changes—at first not so much in marsh biology as in their own experiences. "We talked about getting stuck in the mud more often and falling into cracks that we didn't remember being there last year," Deegan recalls. "At first, I thought it was just because I was getting older, and the new students were inexperienced. But what we were seeing were striking changes in the marsh!"

The marsh grounds became spongier, muddier, and more slippery.

The edges of the tidal creek that let the seawater rise into the marsh began to develop deep cracks. The creek walls could no longer withstand the ocean currents and slid into the incoming tides, piece by piece. Six years later, the marsh's eroded "geography" looked markedly different. But if the extra nitrogen was supposed to cause plant overgrowth, why was this happening instead?

What Deegan discovered is that when stimulated by nitrogen, marsh plants grew taller and leafier, but had fewer, weaker roots. Because there was more food in the water, the plants didn't have to invest in root development. The roots, which normally held the marshland together, were no longer strong enough, and without the roots, the marshland started to fall apart. That's why large chunks of creek walls caved into the tidal creeks and turned into mud. That's why ground that used to be sturdy became spongy and full of holes, which the scientists fell into, literally getting stuck in the mud. As waves battered the soil further, the marsh became muddier, more slippery.

"We discovered that nutrient enrichment was turning the marsh into a *more muddy* place," Deegan says. "Instead of near continuous bands of marsh grass, we now had marsh edges that were lumpy, with large patches of bare mud where marsh grass had been just a few years ago."

Why does this matter?

Walking through a marsh isn't as pleasant as romping in a meadow, strolling through a forest, or bathing on the beach, but coastal marshes are very important. Their collapse has three serious consequences. First, marshes protect the coast from storms and floods. They absorb incoming surges and waves, which sink into the marsh rather than inundating or smashing into houses and other structures. This function is becoming an ever more important one as the climate changes and weather events become more extreme. However, over-fertilized, nitrogen-polluted marshes turn into mud flats that can't absorb incoming seawater. In this era of sea-level rise and increasingly severe hurricanes, the marsh destruction problem is more dangerous than ever.

The second problem is that marshes serve as nurseries for fish and other sea creatures, and when these populations collapse, so do our food sources—no more shrimp cocktails, crunchy crabs, or fried fish that used to feed on marsh dwellers. The TIDE experiment proved that during the first six years of nitrogen dumping, the tidal creeks' fish populations grew, but after the creek walls fell apart, they abruptly collapsed.

The third problem is global in its scope. Healthy coastal marshes are a very efficient carbon sink—more efficient per square inch than forests, thanks to all the thick, heavy grasses and other plants that sprout up, sequestering carbon in their biomass. But if those plants die out, leaving behind dead, muddy banks, all that carbon quite literally goes "puff" into the air. The dead plants rot, decompose, and dissolve into carbon dioxide, the greenhouse gas that's heating up our planet.

The TIDE site on Plum Island allowed several hundred scientists from institutions all over the United States to conduct ecosystem studies. They completed their PhD theses and published over a hundred papers, using the data gathered at that unique testing ground to investigate what happens if we keep overloading marshes with nitrogen. The results weren't pretty. It turned out that plants and fish aren't the only organisms affected. The microbial communities change, too—some become more dormant, others more active, and that alters the overall ecosystem balance, ultimately producing even more carbon dioxide. When pushed to the brink, the marshes can "flip," turning from carbon sinks into carbon emitters—speeding up the dreadful warming cycle and all the evils that come with it.

That possibility becomes painfully clear after we measure marsh respiration at our three Cape Cod sites. As one would expect, the intact Sage Lot had the lowest carbon dioxide emissions, Waquoit Bay had slightly higher readings, and Quashnet Pond had the highest.

The problem is global. Cape Cod's septic tanks aren't the only systems that leach nitrogen into the water nearby. Many traditional sewage systems still do that. Today, America has more than 2 million miles of sewer pipes underneath it, through which it flushes over a trillion gallons of water and waste each year. At waste treatment faculties, this water is extracted, treated to kill pathogens, and then released into lakes, rivers, or the ocean—while still carrying enough nitrogen to cause an overload. New technologies are being added to take some of the nutrients out, but that's not a uniform practice yet. So the Great Sewage Time Bomb still keeps ticking. The marshes will succumb to destruction if we don't change our sanitation systems.

But this is only one half of the nitrogen overloading story. The other half comes with our food—or, rather, with where our food comes *from* in the twenty-first century. The result is what some scientists call the great metabolic rift and others refer to as the redistribution of nutrients on the planet. But no matter what you call it, it's a problem.

The Redistribution of Nutrients on the Planet

The next time you go grocery shopping, take a look at where your food comes from. If you live in a colder climate, most of it isn't local. In the northern United States, your strawberries probably come from California or Florida, your asparagus from Mexico or Chile, and your bananas from Ecuador or Costa Rica. Your salmon probably comes from Alaska, your beef probably originates in Texas, and your pork sausages aren't stuffed by your local butcher, either. Most of the food that gets put on our tables these days is shipped, trucked, flown, and in some cases, even helicoptered to us from far away.

As it grew, your food had to extract nutrients from the soil in which it was planted. Then it was shipped to you—using fossil fuels. Next, you eat the food—and you excrete the unused nutrients, which ultimately end up in a local body of water, probably closer to your house than you think. So we are continuously taking nutrients from some parts of the planet and discharging them in others. The result? Depleted, barren soils in some places and overfertilized, stinky, dying creeks and marshes in others.

Today, farmers in America and beyond spend billions of dollars on synthetic fertilizers, trying to keep their soil fertile year after year. But when we continuously ship corn from South America and strawberries from California to other places—to be eaten and turned into waste—we deplete soils, and then use synthetic industrial fertilizers to replenish their nutrients. What we completely leave out of the equation is that our waste—the potent fertilizer we produce regularly—goes to fertilize all the wrong places: not the farm fields, but our rivers and lakes. And because we don't ship our shit back to where the food came from, this problem will only get worse. Some scholars have called this changing geography of excrement a metabolic rift—meaning that body wastes are generated too far from where the food that fed those bodies comes from.

Canadian epidemiologist David Waltner-Toews calls this situation a "redistribution of nutrients on the planet" that ruins the nutritional balance of ecosystems. "You're taking all this biodiversity out of one ecosystem and creating these piles of shit somewhere else," Waltner-Toews says—hence farm soils turn to dust while waterways suffocate from toxic algal blooms and marshes fall apart. As a result, our continuously depleted farmlands require more and more fertilizer to produce food. With our changing climate, unpredictable weather, droughts, floods, and heat waves, barren soils only add to our already pressing food security problem. We solve this problem by putting in more fertilizer, but we don't have a good way of trapping the excess of nutrients on the other end. Instead, we release nitrogen-rich effluent into the water while the remaining biosolids, aka sludge, are burned or put in landfills, rather than into the fields.

Whatever you call it, the link between humans taking from and giving back to Mother Nature is clearly broken.

There is no doubt that sewage disposal methods in modern industrialized countries are more hygienic overall than those of eighteenth-century London or Paris. We no longer have cholera epidemics, and our riverbanks don't emit a nauseating stink. But we haven't gotten anywhere near the recycling prowess the Chinese, Japanese, and Flemish exhibited centuries ago. Today, even those nations have adopted the industrial methods of sewage disposal that have brought with them larger, more global, and more nuanced problems. No matter what humans do, our shit comes back and bites us in the rear. It really does look as if we can't win the waste war, ever.

Can the solution be found in growing food locally as much as possible and supplying city residents with composting bins? Is it worth revisiting the idea of sewage farms, now that we have much better technology and equipment? Or should we pump our sewage into barges and ship it to Florida and South America to scatter on the fruit orchards? If oil tankers can ship oil across the ocean, why can't shit tankers ship shit?

There's no single one-size-fits-all remedy for the problem of human sewage. A setup that would work for a small town in the countryside probably wouldn't work for a large city. A solution that would work in hot, arid weather wouldn't work in a place with long, cold winters. A small, off-the-grid community might devise a genius composting solution that busy residents of an urban metropolis would not have the means to adopt.

However, the latest developments in sewage technology show that it is possible to convert our waste into biogas, fertilizer, and other forms of energy and useful products. Some of these varied methods are being implemented already—in both developing and industrialized countries. Prompted by the Gates Foundation's Reinvent the Toilet challenge, scientists and entrepreneurs all over the globe are finally looking at excrement in the same way as our thriftier ancestors did several centuries ago—as a resource, not as waste.

The goals of a twenty-first-century toilet, in addition to reusing waste, are to remove germs, operate off the grid, and be affordable and pleasant to use. People need to really *want to use it*, rather than head into the bush.

Sound impossible? Such new systems are already operating in different parts of the world, some in pilot mode and others in full production. Scientists, innovators, and ecologists are making it happen. The rest of us just need to stop holding our noses and dive in.

CHAPTER 7

LOOWATT, A LOO THAT TURNS WASTE INTO WATTS

Clad in blue overalls and tall, heavy rubber boots, sanitation engineer Edonal Razanadrakoto pushes a clunky metal cart onto a narrow street in Antananarivo, nicknamed Tana, his hometown and the capital of Madagascar. In the Western world, Razanadrakoto's work title would mean an office job with a desk and a computer, overseeing operations or redrawing pipe maps for an upgrade, with perhaps some occasional open-air field trips to check on aging equipment.

Not so in Madagascar. So few of the country's sanitation activities take place indoors that Razanadrakoto's job is an entirely outdoor endeavor that includes a very different set of tasks. Most of it has to be done early in the day, before the scorching sun makes it too hard to push his cart through the unpaved, bumpy streets. I don a Panama hat, rub on some sunscreen, and follow him.

Like any large city, Tana has to deal with tons of human waste daily. But unlike its Western counterparts, this urban metropolis of 34 square miles and nearly 1.5 million people has very little piping running underneath it. Many Tana houses are shacks built with bricks made from red mud, with neither running water nor sewage. The poorer Malagasies carry their water in buckets from communal pumps. The middle-class ones enjoy running water, but often only outdoors—with sinks stuck under trees in their yards. Once used,

The original version of this story, "Reinventing the Toilet," appeared in *Mosaic Science*, June 19, 2017.

that water trickles through makeshift grooves into the street gutters. Those gutters—simple open trenches dug in the dirt, akin to the Minoan and Harappan drains of 4,000 years ago—gather up all kinds of filth, and often stink and overflow.

The toilet situation stinks even worse. Only 2 percent of Tana's residents have flushing toilets at home. The rest conduct their business in outside latrines—little closet-like structures with wooden doors, built over pits about six feet deep. The wealthier Malagasies who have large private yards dig deeper, longer-lasting pits and build sturdier and safer structures. They use thick wood for floorboards to resist the inevitable rot. They run an electrical wire from the house and affix a small lightbulb to the latrine's wall to avoid accidently slipping into the hole at night. Some even splurge on cement to ensure that the floors never cave in, and wash them with scented water after each use. When the latrine fills up, they close up the pit and build another one.

The poorer Malagasies don't have such luxury. Their tiny plots, which often must accommodate chicken coops, storage barns, and rice paddies to feed the family, just don't have the space. In poor neighborhoods, like the one Razanadrakoto works in, multiple families share the same latrine, sometimes far from the house, sometimes with no lighting whatsoever, and sometimes very poorly kept, all of which can be a real hazard. Because they often can't dig new pits due to space constraints, they must empty them regularly. In the Western world, cesspools are evacuated by machinery that leaves no mess behind, but in Madagascar it's done manually—by men with buckets, who don't wear any protective gear and don't take precautions to keep the grounds clean. People who take these jobs are usually poor and can't afford this extra equipment. They just scoop and ladle the filth out the best they can, and wash themselves in the nearest river afterward, right next to women and children doing laundry on the banks. When I ask Razanadrakoto where they offload the sludge, he says, "They just put it somewhere, maybe even in the same river."

But even well-maintained latrines are a time bomb waiting to go off. Madagascar floods regularly. The cyclones, as they are called here, arrive from the Indian Ocean, pummel the coastal villages, and soak settlements inland. Everything gets so drenched that the locals give up umbrellas and footwear—walking through swampy roads is easier barefoot, and you don't lose shoes in the mud. During the rainy seasons, roads turn into mud flats and yards into shallow ponds, so that even city residents can easily grow the delicious red Malagasy rice in their yard paddies, harvesting enough to feed the family year-round.

Flooded paddies are a good thing, but flooded latrines are not. When that happens, the filth rises up to the brim and then slowly flows over, oozing out into the yards, down the streets, and into people's living rooms. In eighteenth-century London, the heavy rains may have indeed cleaned the sewers, washing the filth down to the river, but in twenty-first-century Antananarivo, it percolates right next to the houses, with chickens pecking at it and kids playing in it. Diarrhea from fecal contamination is so common that it's one of the biggest fears the locals speak about. You can never be too prepared for it, they say. If diarrhea hits at night, one may not make it to a latrine, especially if it's not in your own yard, but far away. And if you're rushing, you better be careful to avoid slipping on its slimy floor.

The fluctuating water supply makes it difficult to build a steady, reliable sewage system. Even if Madagascar's government decided to invest in building a centralized underground sewage system, it would be not only expensive, but also prone to frequent spillovers. Even in the Western world, despite the massive pipes and drainage systems, downpours often result in sewage overflows. And in Antananarivo, which is located inland, not on the coast, that flow has nowhere to go. Right now, the sewage effluent from the existing flush toilets ends up in the city's lake, which is slowly dying, choked by overgrown algae and lack of oxygen. It also smells worse than the gutters because it's stagnant and isn't flushed by the rains.

But if water doesn't work to remove sewage in Madagascar, can something else do it? Can we come up with workable alternatives in the twenty-first century?

The company Razanadrakoto works for did exactly that. Named Loowatt, this small but ambitious startup took water—that unruly and unreliable transport medium—out of the process and replaced with something far more dependable and trustworthy: people like Razanadrakoto. His duty is to remove excrement from people's homes—like the Chinese *fenfu*, but in a somewhat upgraded fashion—and deliver it to a processing facility where it gets converted into fertilizer and electricity. We may think of his job as a lowly one, but he doesn't. On the contrary, he's proud of his vocation—just as the Japanese *shimogoe* collectors were proud of theirs. "I'm performing an important service for people," he tells me. "We need more services like this in Tana."

As Razanadrakoto pushes his way into and out of potholes, two white buckets reserved for the waste collection bob up and down on the cart. When we hit a particularly bad spot where recent downpours have carved deep ridges in the hardened mud, Anselme Andriamahavita, Loowatt's manager, who oversees the operations and who joined us on the walk, holds the buckets so they won't fall. And then we continue uphill.

At first, the route is relatively empty, but as we get closer to the neighborhood, it gets more animated and chaotic. Adults and kids carry water from the communal pumps, elegantly balancing big yellow buckets on their heads; dogs and chickens dart around; cyclists drive to market on rattling bikes; and eventually cars, too, join the melee. There are no roads or sidewalks here, just a dry clay path that becomes dust in summers and red goo in rainy seasons. As we walk, we get more than a few curious stares from the crowds. Few Westerners set foot in this poverty-stricken part of Tana, so I'm a spectacle.

We reach the home of today's first customer, Laurencia Alix Ravoniharisoa, a 29-year-old mom with elaborate braids and a timid smile. When Andriamahavita asks her in soft Malagasy if I can watch how Razanadrakoto services her toilet, she snickers shyly, but nods and lets me follow the Loowatt crew into her yard. As we walk to her Loowatt cabin, woven from recycled green and white plastic and blending into the lush green trees, I sneak a peak at her old latrine. The little dome at the back of the properly is remarkably clean and doesn't smell too bad. But just a few inches below the hole, I can clearly see a sordid gray goo with a few lazy air bubbles slowly bursting at the top—a clear testament to the country's sanitation challenges. There's no doubt it will bubble all the way up to the top at the next downpour.

In contrast, Loowatt's toilet is impervious to floods. At first glance, the inside of the cabin isn't that different from a modest Western bathroom. There's a little throne with a seat, not unlike my own at home. There's even a lever to tug on, but it doesn't release any water when pulled. Instead, it engages a biodegradable bag underneath the seat. When users pull the lever, the bag moves inside with a faint rustling sound not unlike the mango trees above—and seals the waste in—out of sight and out of mind. The bag lasts for about seven days, depending on the number of people using the toilet. Loowatt's waste collection team shows up once a week to replace it. If the bag fills up faster, and a "FULL" sticker pops up, the owners text Loowatt to swing by sooner.

Wearing latex gloves, Razanadrakoto opens the toilet's "pedestal," removes a white bag, and tosses it into one of the white buckets he brought along. Then he inserts a new, empty bag, and voilà! Loowatt's waterless toilet is back in service. As I diligently film the process with my phone from the best angles I can get, Laurencia stares at me in total wonder. So does Laurencia's daughter Nyaro, peeking out from behind her back. I understand. If someone tried to film me flushing my Kohler toilet at home, I'd think they were nuts, too.

While Razanadrakoto wraps up, I ask Laurencia a few questions and Andriamahavita translates. Still giggling shyly, Laurencia talks in rapid Malagasy, and I catch the word *tsara* a few times. By now I have learned a few words, and *tsara* is an important one. It means good, healthy, or well, which, given the state of personal hygiene here, is always a concern. The new toilet is *tsara*, Laurencia says. It's clean, it doesn't smell bad, and it doesn't overflow, like the old one in the back of the property. "Soon my daughter will use it too!" she proudly says at the end.

As we walk over to the next house, I hop over a street gutter that carries away the used household water. This one, however, is brimming with some thick, oozing glop. It looks so uniformly gray and gluey that I can't even tell whether it's kitchen slop or a flow from an overfilled latrine. It stinks, but I am becoming so immune to the smells that I can't distinguish them. I carefully walk around it so my shoes don't land anywhere near it.

FIGURE 3. The Loowatt waste collection crew making its daily rounds in Antananarivo, Madagascar's capital. The waste is collected in·biodegradable bags, which are placed in closed-lid containers and manually loaded into a digester to be converted into fertilizer and biogas. In places where water supply is unreliable, container-based sanitation is a workable alternative. CREDIT: DENNIS CIERI

Gloria Razafindeamiza greets us in a small path between her house and her Loowatt cabin, with her three-year-old daughter in her arms. On a small stoop not far from the dreadful ditch, a pot atop a charcoal burner is cooking *vary*—boiled rice, which locals eat for breakfast, lunch, and dinner. Gloria is a tenant, Andriamahavita translates, so she had to share her latrine with other families, who weren't good at keeping it clean. She works for the Ministry of Health, and she decided to buy Loowatt service because she wanted more sanitary living conditions. "The toilet was far, and it was dirty a lot, and when you have to run there at night, in the dark, especially when diarrhea comes up, it's really awful," she reveals. But with this toilet she feels safe. It's right outside her house, and only her family uses it. "My daughter doesn't use it yet, but she will soon," she tells me as proudly as her neighbor.

As we chat, I can't help but ask if sitting rather than squatting makes it better, too. Gloria blushes and cracks up. "Ever since we got this toilet, everyone's taking their time," she giggles. "With the old toilet, you were in and out, you just didn't want to be in there, but with this one people are enjoying themselves." She pauses for a second, deliberating whether to reveal another personal tidbit to a stranger, but then goes for it. "My husband disappears in there with his phone for half an hour," she says, laughing. "I don't know what he does in there, I think he goes on Facebook."

I tell Gloria that men in my family, and perhaps even all over the world, do exactly the same thing, and we all share a laugh. Then I summon my meager Malagasy vocabulary as best I can and point at the cabin. "*Tsara, eny*?" I ask. Gloria's face lights up at my language proficiency. "*Tsara, tsara*," she nods enthusiastically. "*Eny!*"—yes.

Shop owner Eleanor Rartjarasoaniony across the street also describes her Loowatt toilet as *tsara*. "We had so much trouble with the old latrine, it filled up every six months and we had to find people to empty it," she complains. "And when it rains, the filth comes up and we are afraid to fall ill."

While Eleanor's chickens hop around, pecking at everything resembling food, including dog shit and my shoes, she explains the economics of the Loowatt solution to me. Residents pay a $20 deposit for Loowatt's toilet, and a $5 monthly service fee that includes waste removal and fixing anything that breaks. For Malagasy families, many of who live on a dollar or two a day, this isn't cheap. But emptying latrines costs money, too, Eleanor says. "It may even be more expensive, if you have to do it often," she says. "And we did it a lot because we shared our latrine with our tenants, and it's seven people altogether." At the end of her speech, Eleanor proudly makes the same familiar statement as did every mom before her. "Even my eight-year old son can use it."

I am a bit baffled by this pride—what's so special about a child being able to use a toilet? Isn't that something a kid learns by the age of three, if not younger? The Malagasy kids are so mature and self-sufficient—they walk alone on streets that don't have sidewalks, they carry heavy buckets of water on their heads, and they shoo hungry stray dogs from the wandering chickens. Why are their mothers so proud of their toilet-using prowess?

As we leave the shop, I sneak up to Andriamahavita and ask. His answer makes me gasp. Using latrines is dangerous for kids, he explains. The little ones sometimes fall through. Sometimes wood planks crack and break. And the pits are deep, one minute your child is there, the next she's gone. If parents can't find their child, they check the latrine. Every Malagasy mother is terrified of that. As I picture this horrific image, I begin to remember that in my own childhood, our neighbors who had latrines rather than septic tanks were scared of the same thing.

The thought that Madagascar's sidewalk-less streets are less dangerous than house toilets preoccupies me for the entire time it takes us to finish the waste collection route and head back to "the base." There, I observe what happens to all this loot from the loos and how it gets converted to watts.

How Loowatt Got Its Name

In 2000, a young rookie reporter, Virginia Gardiner, went to work at the San Francisco office of an architecture and design magazine called *Dwell*. A recent graduate of Stanford University's Comparative Literature Program, she was the youngest on *Dwell*'s editorial team, so she got to cover the not-so-illustrious realm of household amenities. "Nobody else wanted to go to the kitchen and bath industry shows, so I had to go," Gardiner says. Yet for her, it was a dream job, because she got to travel and see the endless variations on bathroom and toilet designs. Soon she was realizing how symbolic our porcelain thrones were, how much drinkable water was flushed down the drain daily, and how wasteful American bathroom culture was overall. "The first article I wrote for the magazine was about toilets—about how and why they don't change. And I was really grossed out by the culture of consumption of the bath products," Gardiner recalls. She did this for eight years, and then decided to put what she'd learned into practice. "I went to London to get a master's degree in industrial design, and I did my thesis on toilets."

She wanted to build a waterless urban loo that used no energy and turned waste into something useful. She began by building a composting toilet in which she regularly fed poop to a bunch of earthworms, mimicking what normally happens in nature. The method worked amazingly well. "You would not believe how much poop would be transformed in just a few days," she says. "It was as if it had never been there." The next step was to create a miniature anaerobic digester, which used bacteria that broke down the waste and produced methane—once again, just as they do in nature. That methane could be burned to power a generator and make electricity. It was simple and efficient.

Gardiner explained the concept in a video, in which she also showed a toilet made from poo—or rather, from the composted soil-like biomass baked into a pedestal form. The video went viral. Gardiner embarked on a fundraising campaign, titled "Let's turn shit into a commodity!" Shortly afterward, the Gates Foundation issued its Reinvent the Toilet Challenge. Gardiner's design matched the foundation's requirements, and the award money helped her develop Loowatt into a commercial enterprise. Her first business investor was a Canadian expat living in Madagascar who saw the video. And that's what gave Loowatt a start.

Enter the Biodigester

Loowatt's waste processing site occupies a mere 4,000 square feet on top of a cliff overlooking Tana. There are no big buildings, no offices, and no glowing computer screens keeping watch on the daily operations. But there are pipes—lots of pipes—connecting some strange-looking machinery to a tank, over which floats an enormous balloon-like vinyl bag. All the bags Razanadrakoto and his team collect from their customers arrive here.

Like collecting the bags, loading them into the processing equipment is a manual task. Because the machinery breaks the bags as it chews them up, their contents may spew into the air, so sanitation worker Tojoniaina Andriambololona puts on a face mask and hands me one, too. "Please put this on," he says. "We consider this area contaminated."

Mask and gloves on, Andriambololona removes the white bags

from the buckets and loads them into the waste-extraction machine set up inside a tent-like structure. I hold my nose through the mask, but the loudly clanking apparatus lets out surprisingly little odor, perhaps because the waste remains in the bags, which are made from some kind of starch that will fall apart in the biodigester.

Outside the "contaminated" tent, two other workers bring out a bucket of food leftovers—a mix of rice and vegetables from local eateries—which they load into another machine that looks like an oversized meat grinder. It turns out to be exactly that. While one of the workers sloshes the glop into it, the other spins the handle—and the rumbling clunker mashes up the leftovers, spitting out grayish slop. Mixing food and excrement makes the bacterial decomposition process more efficient, Andriambololona says.

Next, the mix of poo and goo is pumped into a pre-digester tank and heated to 70°C to fry the pathogens. Loowatt's method uses hot water to pasteurize the mix, but it relies on self-generated energy to heat that water. The freshly "cooked" matter then gets pumped into the biodigester, a tank with a capacity of about 1,000 cubic feet, where anaerobic bacteria slowly break down the waste, converting it into liquid fertilizer and biogas. The biogas, which mostly consists of methane, floats up into the green bag on top of the tank, blowing it up like a huge balloon. When the bag is full, it's bigger than the tank itself. Loowatt workers use the gas to heat the water that kills the pathogens, to charge phones, and to boil *vary* for lunch.

Andriamahavita points at a pipe sticking out of the biodigester's rear end. "This is where we open it and the fertilizer comes out," he says, adding that because hot water has killed the pathogens, it is safe to put in the fields. I can't help but think back to Pullman's sewage system that used steam to propel waste to farms. I wonder if one reason for its success was that the steam killed pathogens, too.

Loowatt is still in a pilot mode, so its output can't compare to that of industrial systems. It converts the waste of about 1,000 people into about six metric tons of liquid fertilizer a month. But six metric tons is still an impressive amount. "Where does it all go?" I ask Andriamahavita. "What do you do with it?"

"Different things," he replies. "Some of it we sell to farmers. And some of it we keep for ourselves. We experiment with it."

"Experiment?" I echo him. "What kind of experiments can you do with recycled poop? Can I see?"

"Sure," Andriamahavita says. And he takes me to Loowatt's research and development lab.

Disinfecting Thyself

Driving in Tana is murder. New York traffic has nothing on Antananarivo's oversaturated, potholed streets, in which one has to constantly maneuver among pedestrians, cyclists, street carts, stray dogs, and runaway chickens. With the exception of the city's center, where straight lines and streetlights maintain a semblance of order, the metropolis is a churning, seething chaos where everyone is simply going about their business. While we sit in traffic, whether because a cart up ahead lost its wheel or because a donkey refused to move, Andriamahavita tells me how Loowatt first got started and where it wants to be a few years from now.

One of the most significant challenges Loowatt first faced was finding a piece of land to put the digester on. To make the manual collection system work, the digester site had to be very close to the neighborhood it served; otherwise, the operation didn't make sense. Matching available plots of land with customers' proximity proved difficult. "Space is a real issue in Antananarivo," Andriamahavita laments. "Nobody wants to give you space for processing waste. People have a hard time understanding why it's even necessary. They are not used to paying for waste removal. They are used to thinking that they can always dig another hole in the ground. But with so many people living in Antananarivo, we don't have enough space to dig more holes." Antananarivo, whose name means "The City of a Thousand," was a marvel of civilization centuries ago, Andriamahavita explains. A thousand people living here had enough land for latrines. A million and a half today do not.

To be an economically viable sanitation company, Loowatt aims to expand to at least 5,000 customers in the next couple of years. Gardiner thinks that Loowatt can reach double that. The biggest challenge isn't the technology, but the manual collection process, she thinks. Without water as a transport medium, it isn't possible to automate that part of the process—which brings us to the same centuries-long conundrum.

But Armel Segretain, Loowatt's operations engineer, who left Europe for Madagascar to implement this unconventional sanitation process, says that waterless technology has some advantages over Western systems: "Ninety-five percent of the sludge in most industrial sanitation plants is water. That water must be filtered out and treated with chemicals before it can be safely released into the environment." Loowatt has very little water in its sludge, except the heated water used to kill the germs. That reduces the size of the operation and increases its efficiency. "It makes the entire setup smaller, cheaper, and easier to operate."

Segretain thinks the logistics of the waste collection can be worked out, so after vetting the process with its 100-toilet pilot, Loowatt is getting ready to scale up. "Our goal is to become the major sanitation provider for the whole city," he reveals. "It will take awhile, something like five to ten years." It is also gearing up to expand to other countries. In fact, Loowatt toilets are already being used in the United Kingdom, albeit in a very different way—at festivals and other outdoor venues. Those toilets make good money, Segretain remarks. "In Madagascar, the most we can charge people for our service is $5 a month," he explains. "For our festival toilets in the United Kingdom, people pay that amount *per use*."

Andriamahavita and I arrive at Loowatt's first toilet site, and I learn that some Malagasy Loowatt clients also pay per use—albeit a lot less that five bucks a clip. A sign on the door details the pricing in this order: *Kaka 100 ariary/Pipi 50 ariary*. Malagasy concepts of number one and number two seem the opposite from ours, but it's not hard to figure out what's what, and I add two more words to my Malagasy lexicon. A quick calculation on my phone reveals that *kaka* would cost me three cents and *pipi* a cent and a half. A family with a few kids subsisting on a dollar a day would have to carefully ration not only their intake but their output, too. I can see them still choosing a bush over Loowatt.

But an elderly man named Felix who mans this operation says that the toilet is popular and the biodigester attached to it fills up. "We get one cubic meter of biogas from it twice a day," he informs me as Andriamahavita translates. "Right now the bag is empty, because we burned it, but at 6 pm it will be full again. And then residents can come and charge their phones." This community is so poor that it barely had any toilets or electricity at all, Andriamahavita adds. Now it has both. The residents use their own organic power to charge their phones, what can be cooler than that? And with that comes Loowatt's R&D facility.

I follow Andriamahavita into the lab. And I gasp in shock.

Pushed against the walls like bookshelves in a library, crates of shit are stacked up all around us. There's shit everywhere—in bins, pots, buckets, and cans. Two men and a woman, their heads covered and their hands gloved, are shoveling it around, checking and measuring something. The piles look and smell more like cow manure than neighborhood latrines, but I still can't quite shake off the shock. Watching excrement removed from toilets in gleaming white bags is one thing. Seeing it towering around you, up close and personal, is another story.

Andriamahavita introduces me to the three workers, who smile broadly, peel off their blue latex gloves, and stretch out their hands to shake mine. I freeze in horror. According to the unwritten rules of journalistic ethics, I must do the handshake. It's a simple act of respect—to my sources and to any human being I meet. But what about the dangerous pathogens lurking in that sewage? I had the immunization shots my travel doctor suggested, but I'm not immune to cholera, intestinal worms, or *E. coli*. So would shaking hands be a requisite part of my job, or an act of utmost stupidity?

Seconds pass, and the three Loowatt workers are still waiting. I draw a deep breath, hold my tape recorder in my left hand—and shake the three hands. I wonder if this is how the people who empty latrines feel when they go down into the pits. I wonder if this is how my grandfather felt when he bent over our septic tank to dip his buckets. I mentally kick myself. I carried a sanitizer bottle with me all day, and now I have stupidly left it in the car.

And then I realize that Andriamahavita is talking about composting. When the processed waste comes out of the biodigester, it can be used as liquid manure, or it can slowly become compost over about four months. "But if you feed it to the worms, you get vermicompost—rich, black, fertile dirt—in half the time," Andriamahavita explains. "Vermicompost goes for a higher price. So yesterday we decided that we will stop making regular compost and only do vermicompost."

As he talks, I realize that I overreacted. The stuff in the crates has already been pasteurized and processed by microbes, and is now being chewed up by worms. That makes it no more pathogenic than the black dirt I shoveled from my grandfather's compost pits into strawberry patches. It certainly does contain germs, and I should thoroughly wash my hands when I leave, but it is far less dangerous than the contents of the pit latrines or street gutters.

As if reading my thoughts—or maybe my face—Andriamahavita snatches a large sanitizer bottle, pumps some out, and passes it over to me. I smear it all over my hands and finally dare to use my right hand to pull out my phone from my pocket and take a few photos.

"Do Loowatt employees get immunization shots against sewage pathogens?" I ask.

"Yes, we all got our immunization shots," Andriamahavita replies.

When we walk out, I step on a small chunk of something brownish-black. It looks like poo on its way to becoming compost—or maybe just some "unprocessed" raw feces from a street gutter. As I try to shake it off, I realize that I will have to pack my shoes into my travel bag at some point—right next to my shirts, underwear, and toothbrush—and the hair on the back of my neck stands up in horror. I don't think I have enough sanitizer to decontaminate my shoes. Neither do I feel like touching them with my bare hands.

I think of how my grandma used to decontaminate my grandpa's work clothes, and ask Andriamahavita, "Can I buy an electric kettle someplace?"

He nods. "There's a street market in the center of Tana. It's on the way. I'll stop there. Just be careful," he warns, "because the market has many pickpockets and you have a very nice cell phone."

An hour later, I walk into my hotel room, kettle in hand. I take off my shoes and put them soles up on my shower floor. When the kettle starts boiling, I pour the gurgling water over their bottoms over and over again, until I'm convinced that no germ could survive this scalding downpour. Then I take a long, soaking shower, covering myself in soap top to bottom. Our Western ways of dealing with shit may not be perfect, but I now understand why we built our sanitation systems the way we did.

Yet, Loowatt's collection and conversion method is ecologically superior to our traditional sanitation systems. It repurposes waste the way Mother Nature does, separating it from us while making it harmless. Has anyone else tried to implement something so simple and smart?

Yes, it turns out—and it's working out quite well.

CHAPTER 8

THE CRAP THAT COOKS YOUR DINNER AND CONTAINER-BASED SANITATION

Once upon a time in 2011, three sanitation specialists, Emily Woods, Andrew Foote, and Chris Quintero, were sitting in their top-floor apartment in Santiago, Chile, discussing the possibility of infecting themselves with the intestinal parasite *Ascaris lumbricoides*. And they were serious about it, too.

A. lumbricoides is a gut worm, about 15 inches in length, that hangs onto the walls of its host's small intestine, sucking out nutrients and shedding over 200,000 eggs a day. When its eggs leave the host's body in feces, they typically reach new hosts by way of unwashed food or hands. Once back in a small intestine, the larva hatches and migrates through the intestinal mucosa into the blood, heading straight for the lungs. Once the little buggers mature, they climb up to the throat, making the host re-swallow them. At that point, *A. lumbricoides* settles in the small intestine for good, grows into an adult worm, and begins to shed its thousands of eggs.

Severe infections manifest themselves in malnutrition and swollen bellies, often very noticeable in little kids in parts of the world with poor sanitation. That's why when ascaris parasites are involved, the discussion usually revolves around their elimination, not around becoming a willing host.

But the sanitation trio had reasons for their unusual idea. They came to Santiago on a business grant from the Chilean government aimed at fostering entrepreneurial spirit. The team wanted to test the effectiveness of a simple and inexpensive method of killing sewage pathogens: a solar latrine. When heated to and kept at 42°C for a few hours, all pathogens die. If someone were to attach a simple parabolic mirror to the back of a latrine located in a hot, sunny climate, and were to direct the sun's energy onto the sludge, after a few hours of that heat, all germs would be dead. Unlike chemicals, which are expensive and not always available in the developing world, the sun predictably shines for long enough in many places to render feces germ-free.

These sewage pathogens come in many shapes and sizes, but the team was particularly interested in *A. lumbricoides* because it serves as an indicator organism for pathogen elimination. If you kill, or "deactivate," the eggs of *A. lumbricoides* in fecal sludge, you know all the other pathogens in there are dead, too. In many places in the world, particularly in sub-Saharan Africa and parts of Asia, *A. lumbricoides* infections are nearly universal. The worms dwell in many people, and their eggs are likely to be swimming in every pit latrine. But Chile wasn't one of those places. *A. lumbricoides* isn't endemic to Chile.

So when the team had set up their testing equipment—a simple mirror attached to a black metal bucket—and was ready to start the efficiency testing, they suddenly faced an unforeseen problem. Where would they get a regular supply of *A. lumbricoides* eggs? Finding infected people in Santiago didn't look promising.

So the team pondered the idea of swallowing the parasite so they would start regularly producing its eggs, in order to test their solar latrines on their own poop. "We needed to do it in a repeatable, consistent way," explains Quintero. "We didn't want to start selling someone our latrine that may or may not work correctly all the time. So we spent a lot of time to see if we can infect ourselves and a couple of other people on the team and produce a regular supply."

This radical idea isn't really that scary as one may think, he adds, because *A. lumbricoides* isn't that bad of a parasite. In many people, the infection produces no symptoms, and while the worm does suck out some nutrients, it takes time for severe effects to develop. "And it's very treatable with modern drugs," Quintero adds, "so we would only have to do this temporarily."

But the idea had other problems. First, the team would have to procure an initial supply of eggs. The worms take a few months to mature before they start laying eggs, and the trio's grant had a time limit. There were also some logistical problems. The team's lab, where the buckets and mirrors were set up, was not near their apartment. Having to run from the apartment to the lab to use the latrines was not an ideal situation. The team considered putting their solar latrines on the rooftop of their apartment building, but the plan drew little enthusiasm from the building management. The sanitation team may have been willing to take risks in the name of science, but others were not.

The team had to find another creative solution for their testing—and they did. They discovered that the Chilean dog population commonly carried a parasite very similar to *A. lumbricoides* that behaved in much the same way. So, instead of human feces, the team used dog poop—which proved to be easily accessible. "Emily and Andrew went to dog shelters, collected the infected matter and tested it," Quintero recalls. "It had the same characteristics."

The tests worked as a proof of concept: the mirrors consistently heated up the fecal matter and fried the eggs. "We were so excited," Woods recalls. "Being young and fresh out of college, we were like, this is the solution that's going to save sanitation." They wanted to take their model to poorer places, like sub-Saharan Africa, and launch a solar latrine business.

Their initial thought was that people would buy their toilet setup and do "self-treatment" of their feces, which would help curb infections and diarrheal disease. But they quickly realized they many people simply didn't have the money to pay for it. Among their other daily problems and expenses, treating waste for pathogens wasn't a high priority. Besides, treating it didn't eliminate it—the owners would still have to find ways to dispose of the treated waste. So even though the process worked in principle, the business model of selling self-treating toilets was problematic. The team would have to find another opportunity to put their invention to good use.

What didn't quite work out during their boot camp in South America took hold a year later, halfway across the globe—in Kenya.

The Flying Toilets of Kenya and the Business Side of Poop

Striding through the urban slums in Nairobi, Kenya's capital, requires special treading skills. One has to avoid stepping onto one of the plastic bags—usually tied and slightly puffy—that litter the roadways. It's easier said than done—in some places, the roads are covered with bags of different sizes and colors. Step on one accidentally, and the thin plastic bag will probably rip, squirting your shoes with the black or brown gunk inside. It's human waste.

The largest informal settlement in East Africa, with over 600,000 people, the Kibera slum is extremely overcrowded. Stuck together with barely any space left, the tiny shacks people live in have no places to dig even the simplest pit latrines. But the people still have to relieve themselves somewhere. So they go in the plastic bags— and then throw them away by the roadside, into ditches, or just as far away as possible. The bag phenomenon has a name. It's called "Kenya's flying toilets."

In addition to flying, the bags migrate. They drift, pushed around by street dogs and people. They float in the streams that form on the muddy roads after heavy rains and gush down the streets, piling up in new locations until the next downpour. They also rip, spitting their contents onto the streets and into streams and gutters. During certain times, parts of Kibera get so littered with these bags that they become impossible to avoid.

The flying toilet phenomenon is specific to Kibera. In the rest of the country, most people use latrines, which come with problems similar to those in Madagascar. Latrines overflow during rainy seasons, flooding roads and homes with excrement, sometimes causing such terrible fecal contamination that schools and hospitals shut down to avoid disease. And just as they do everywhere else, the latrines fill up and must be emptied, either manually or with imperfect machinery that leaves a mess behind.

None of this looked like the right sanitation approach to Woods, who came to Kenya in 2012 as a consulting sanitation engineer, helping an NGO build 900 pit latrines in the Naivasha area, a quickly developing region. About 47 miles northwest of Nairobi, the area was experiencing an economic boom. The flower farm industry was blossoming, the geothermal power plant near Lake Naivasha needed workers, and the local wildlife resorts were bringing a stream of international tourists, but there was little sewage infrastructure to support the population surge.

Woods was asked to design better and cheaper pit latrines, but when she did her analysis, she realized that merely building 900 latrines would be a stopgap measure—a patch on the existing problem rather than a solution. With no proper way to treat or dispose of the sludge, the latrines would simply fill up. "You dig 900 pit latrines and then three years later you have 900 more to dig," she explains. "I felt like I wasn't actually doing anything to solve the problem."

That was when she first thought of creating a business that would turn the human waste into something valuable—something people would want to pay for. She reached out to Foote and asked him to come to Naivasha to brainstorm a solution for sewage repurposing. They had tested their solar sewage treatment method in Santiago the year before, so all they needed was to figure out was a way to put all that "safe poop" to good use.

Together, they began to ponder a for-profit sanitation company that would collect sludge from low-income households and convert it into a product that would be in high demand throughout the country. In Kenya, where average family income is about 10,000 Kenyan shillings a month—less than US$100—that meant that their product would have to be very affordable. To put it plainly, Woods and Foote would have to convert human poop into something many humans would buy and use often.

Could they sell poop back to the people who made it?

The duo approached the idea from a hard-core business angle. "We looked at every potential reuse of feces and did the analysis: How can we make the most money per kilogram of raw feces?" Woods says. Some methods were ruled out from the start. For example, converting sewage into fertilizer and biogas wasn't going to be a viable business. You have to look at efficiency, Woods explains. A kilogram of feces yields only a very small amount of biogas. And once you produce it, you have no means of distributing it to your customers because there's no pipeline infrastructure. It doesn't fit the cultural norms either: Kenyans don't cook with biogas; they use charcoal and wood.

If they went the fertilizer route, the returns didn't look promising either. Converting feces to compost, or soil, means that your starting product breaks down and decomposes, Woods explains. "That means that whatever mass you started with will get smaller and smaller and smaller," she points out. "So you will have less of the final product than what you started with—and therefore less to sell." Plus, you'll be stuck with having to deliver your product to farmers. That product will also have to be something that not only farmers, but the vast majority of Kenyans, will use. What's more, producing fertilizer takes a long time, and the feedback from customers is slow. "If someone buys soil from you for the crops this season, you are not going to know if they liked it or not until a year later," Woods says. A fledgling product needed more flexibility.

When Woods and Foote ran their analysis, they realized that solid fuel might be the answer. In a country strapped for energy with rising demand, fuel was one commodity everyone needed—every household, hospital, school, and factory. And unlike biogas, solid fuel could be distributed easily and cheaply, without pipes or pressurized containers.

Over 80 percent of Kenyan urban households cook with charcoal, which is made by cutting trees and charring wood in low-oxygen conditions—that is, by carbonizing it. Since 2004, the country's charcoal use has increased more than 50 percent, and prices have nearly tripled. There are sustainability regulations that govern the process of wood harvesting, but the demand is so great that illegal markets blossom—meaning the majority of charcoal made in Kenya doesn't comply with sustainability laws.

The other type of fuel that's popular in Kenya is firewood. Charcoal can smolder for a long time, generating the long-lasting, steady heat necessary to boil water or simmer a stew—which is why families seek it for cooking. Firewood generates a lot of heat very fast, but not for long. That doesn't work for cooking dinner, but it's good for certain businesses, Woods explains. For example, Kenya's tea companies use firewood to dry tea leaves quickly. (Gone are the days when the leaves were slowly dried under the hot sun—that's inefficient, weather-dependent, and requires large plots of land, which not every company has.) Some tea factories process hundreds of tons of tea leaves a day—and that requires an extensive amount of firewood. "They have ridiculously high biomass fuel demand," Woods says, so they would make consistent and reliable customers.

If Woods and Foote could figure out how to convert feces into a safe and efficient fuel that fit the bill and didn't look or smell like shit, they could tap into a very promising market. In Naivasha, charcoal sells for about $0.30 a kilogram, and the city's 100,000 residents use it every day. The market in nearby Nairobi is much greater—over $3 million. "The market in Kenya is huge—3.2 billion US dollars," Woods says. "And it's everywhere."

That meant they could indeed sell poop back to the people who made it. As long as it didn't look like poop and burned well.

Some forms of feces—such as cow or elephant dung—burn easily. That's because these animals consume a lot of grass and leaves, so they excrete a lot of undigested fiber, which burns well. Human poop has a slightly different composition, but it contains a lot of lignin—a complex organic polymer that comes from the cell walls of plants. Because of lignin, human poop transforms into a sticky substance when heated. If you throw more burnable fiber into that sticky mix, the result can be shaped and dried into fuel briquettes.

Luckily, Kenya produces various fiber-rich farming leftovers. Naivasha grows and ships roses all over Europe, so the plants' leftovers are abundant and inexpensive. Woods and Foote carbonized rose waste by a process akin to charcoal production, ground it into powder, and added it to heat-treated feces. They fed the resulting mix into a machine that shaped it into round balls. After three days of drying, the balls burned like charcoal.

For steady briquette production, Woods and Foote needed a consistent input stream. So they supplied customers with container-based toilets and put together a waste collection team that regularly gathers the contents, much as the Loowatt service does. However, their toilet's inner workings are different. There's no biodegradable bag; instead, the toilet has two separate parts—one for urine (called a diverter) and one for feces—because keeping liquid out of the sludge simplifies its processing. That potty setup requires some getting used to—the sitting down and "aiming" takes a bit of training for the user, but eventually everyone learns to position themselves, Woods says. "A lot of people are nervous the first time they use it, because they worry they are going to miss," she explains, "but after they've used it for a couple of weeks, most of them are comfortable."

Woods and Foote's company, which they named Sanivation, had a pilot project running in 2014. A year later, they finished building their first waste treatment facility and have been in business ever since. Their collection teams visit customers twice a week. Wearing gloves and masks, the servicers swap tightly closed containers without any exposure to feces—a process far more hygienic than emptying latrines, whether manually or with machinery. In Kenya, there are exhauster trucks that suck up latrine contents, but they create a lot more mess, Woods says. "There's a lot of splashing, a lot of smell, and at the end the truck has to dump it somewhere, so you are much more in contact with feces than you are in a container-based situation," she explains. "The process of putting a lid on a bucket and then grabbing that container and putting in an empty one is one of the most hygienic options of waste transfer. It's a lot less nasty than working an exhauster truck and way better than manual emptying."

Over time, Woods and Foote had to make many adjustments to their operations. Originally, they were charging 600 Kenyan shillings a month for their service—about 6 percent of an average household's income—which proved too expensive for many families. When they dropped the price to 200 shillings, a lot more customers signed up. "It still costs us 600 shillings per month to do the service," Woods says—so they had to find other ways to balance the books.

They managed to carve out some savings by creatively tweaking the fuel formula. Rose waste proved labor-intensive to obtain. It required workers to gather the leftovers scattered in the fields, which took time and effort. "We used it earlier because rose waste was everywhere and we thought it would be the cheapest way to do it," Woods explains, "but the cost of collecting was too high—the waste is just out there in the fields, it's hard to collect." The cheapest input stream turned out to be charcoal dust, a leftover from charcoal production. "Any big bag of charcoal has anywhere from 10 to 20 percent of it—small bits and pieces that can't be used in a normal stove," Woods says. "We have it collected and delivered to our site and we add it to our poo binder. It's the cheapest way to do it right now."

Sanivation also tapped into the firewood market, essentially making logs from poop. Here the input stream is sawdust, which comes from a grinding mill—one centralized spot, so it's easy to gather and transport. Sawdust isn't carbonized, so the logs made with it burn much like regular firewood, generating a high temperature very fast. The demand from factories is enormous, Woods says. "We started offering our non-carbonized logs as a product four months ago and we just sold 50 tons last month. And we can sell more, we are only limited by how much we can produce."

Woods and Foote plan to scale the company up and serve 1 million people soon. They also plan to start accepting exhauster truck contents, which usually get dumped somewhere, as their new high-volume waste input stream. Besides Naivasha, they also offer their toilet service in Kakuma, a refugee camp of about 60,000 dwellers in northwestern Kenya. In the camp, their business runs pretty much full circle—the residents cook on their own poop, with some added fiber thrown in. "In Kakuma we sell briquettes to the same people we collect waste from," Woods says. "It's kind of neat how it works."

Are We Ready to Bring Back Chamber Pots?

Besides their obviously innovative methods of sewage upcycling, Loowatt and Sanivation have introduced (or perhaps reintroduced) another curious concept: container-based sanitation, or CBS. With the CBS approach, flushing toilets simply don't exist, because all the waste is gathered in a container underneath the toilet seat, eliminating the need for water. Only, in contrast to the *Tout-à-la-Rue* method, these containers aren't emptied onto the heads of the unsuspecting strangers.

These two companies aren't the only ones to arrive at this solution. There is a CBS alliance formed by a half dozen container-based sanitation companies that operate in different parts of the world. You can think of them as the CBS Six. In Haiti, it's SOIL, an acronym for *S*ustainable *O*rganic *I*ntegrated *L*ivelihoods. The company collects toilet output from families and converts it into nutrient-rich soil using thermophilic composting, which engages bacteria that like heat; the resulting mix helps restore Haiti's depleted lands. In Ghana, Clean Team produces fertilizer and energy as its final products. In Lima, Peru, one of the driest cities on earth, which barely gets any rain at all, it's X-Runner. And another CBS company, Sanergy, which operates its Fresh Life Toilets (FLTs) in Nairobi's slums, has perhaps the most unusual approach to the problem. It feeds the feces to the larvae of black soldier flies. Once the maggots are fattened up, they are boiled, dried in the sun, and milled into a protein-rich animal feed. It's remarkable how versatile a resource poo can be.

Despite the variety of waste conversion methods, all CBS companies have certain things in common. A CBS toilet must capture its input in a sealed container, eliminating contact with feces for both the users and the servicers. Some CBS toilets separate urine and excrement to reduce the waste's volume and weight for easier repurposing. When full, the containers are sealed, exchanged for empty ones, and delivered to a processing facility without contaminating their surroundings or the air. The containers are disinfected and reused—and so is the waste. It's treated and recycled into a usable form—typically some form of fertilizer or energy, both of which are, needless to say, vital.

Freeing toilets from the constraints of water can be amazingly liberating. The toilets are no longer soldered to an extensive pipe system, so they are portable and can be moved around the house or between different houses. They can be made locally with recycled materials, shipped easily, and installed with minimal costs. They fit perfectly into densely populated low-income neighborhoods or settlements, regions that flood quickly, and places where building water-based sewage systems isn't feasible or appropriate. Even in areas where water-based sewage systems can be built, CBS systems represent a much more affordable solution for low-income families than running pipes and connecting them to centralized systems. CBS sanitation is also a cost-effective option for financially strapped municipalities— it's substantially cheaper than running pipes and redirecting water, and also far less resource and energy intensive. It also works well in refugee camps or disaster areas, as well as in hospital settings, where CBS toilets can easily serve as bedside potties—especially in hospitals where the only other option is an outside latrine.

This type of sanitation may appear very primitive compared with the shiny porcelain johns of the West, but in fact CBS toilets are in many ways superior. And they are far more sustainable and easier on the planet and its resources. So far, Western countries haven't advanced very far in sewage upcycling, and as we saw in chapter 6, this is taking a substantial toll on the environment.

If CBS sanitation is so efficient, can the rest of the world adapt the concept? Woods thinks it's possible. The flush toilet became the symbol of sanitation partly because in some parts of the world in the twentieth century, water seemed cheap. But as water becomes a more precious resource, not all places that currently flush may be able to continue. So sanitation experts are realizing—perhaps for the first time since the beginning of the sewer system era—that water-based systems aren't the ultimate approach.

Woods believes that water-based sanitation isn't sustainable because of its high infrastructure, maintenance, energy, and water costs. The global long-term sanitation future may be dry. That means that some 100 years from now, the sewer systems of the United States and Europe will look drastically different. Or they will be replaced by something else. Woods thinks that in a few years, Sanivation's approach may be the model for unsewered waste processing. What Sanivation and other CBS companies are doing with their dry, container-based toilets may be the beginning of the clean, waterless future of humankind's waste.

Will that future involve containers? Will there be a service truck buzzing around New York City streets picking up sealed buckets along with garbage and the recycling we so diligently put out the door today? It wouldn't be all that different from the old Chinese,

Japanese, and Flemish systems, which served their populations well. But would people in the Western world, accustomed to the convenience of the flush, give up their comforts and start putting malodorous buckets out the door for the pickup crews? And wouldn't that go against progress?

If you can no longer flush because there's no water in the pipes, a modernized version of a chamber pot may suddenly look quite appealing. After all, Westerners have learned to separate their garbage into various waste streams, such as compost, paper, and plastic, so placing one more bin outside for a weekly pickup wouldn't be much of a chore. So perhaps CBS sanitation can indeed be done on a more modern scale with smart, sealed, ergonomic containers, perhaps even with smiley poop emojis on the sides.

Does this mean that we have come full circle, arriving back at the wisdom of our ancestors? It looks that way to me. But with our current knowledge of health, disease, and bacteriology, we can do it better. With some effort, we may even forgo the container part and turn to toilets that can covert our waste into fertilizer and fuel on the spot.

Sound like a distant utopian future? It's no longer a future, but a real-time present. You can buy a toilet like that right now, if you want to try it at home. And that toilet has a story to tell.

CHAPTER 9

HOMEBIOGAS: YOUR PERSONAL DIGESTER IN A BOX

One day in 2006, a soul-searching, globe-trotting student climbed a tall mountain in rural India and had a life-changing experience. His name was Yair Teller, he was 27, and he was studying marine science and water ecology at the Ruppin Academic Center in Israel, but he wasn't sure what he wanted to do with that degree. Would he be ever able to apply it to anything useful? And if not, what was the point of spending time on it?

So Teller took a year off and went to see the world. While traveling in southern India, he found himself standing on the mountaintop and looking at a farm. The farmer's family invited him for a cup of chai. As they put on the pot to boil the water, Teller watched them light a blue flame on a stovetop. He was shocked. It was the first smokeless kitchen he had seen in India.

When it comes to cooking, India has a massive problem. About two-thirds of its 1.2 billion people still cook on open fires, and many still forage for fuel—dried cow dung, wood, or charcoal. Women, who are usually charged with gathering wood, carry it home on their backs. Men cut down trees, adding to deforestation. But the fires themselves are the worst offenders. They produce plumes of smoke and soot—also called black carbon—which is extremely damaging to humans and the planet. When that black carbon floats over the Himalayas, it coats white ice with black soot, which absorbs the sun's heat more intensely and intensifies glacial melting. The soot also settles in people's lungs, causing asthma, cancer, and a slew of other disorders.

Some data show that worldwide, nearly 4 million people die each year from exposure to these particulates[1]—more than from malaria, HIV, and tuberculosis combined. Women and girls, along with all the little children in their care, get the lion's share of that exposure because they tend the cooking fires.

"Where do you get the gas for this beautiful blue flame?" Teller asked his hosts in awe. He was familiar with India's cooking issues and knew there were no gas pipes running up the mountain. He knew that regular gas delivery was an unlikely scenario. There must've been a different source for the clean, smokeless fuel they were burning.

That source was the family's farm itself. Or, to be precise, its waste.

The farmer took Teller outside and showed him a covered concrete pit and a pipe that stretched from it into the kitchen. "We have a lot of cows," he explained. "During the day they graze outside and at night they come home. Every morning we sweep their dung into this pit and we leave it here to ferment. That gives us gas and liquid fertilizer. We use the gas to make that beautiful blue flame you saw, and we pour the fertilizer onto our fields to grow food. We also use it to grow these white flowers called kala, which we sell in the market."

The farm got its biodigester as part of India's program to create sources of energy in rural locales. The government helped the villagers build their digesters and split the cost fifty-fifty.

Teller was amazed. The design was simple, yet efficient. It generated clean fuel, it fertilized fields, and it created additional income for the family. As the farmer proudly explained to Teller the inner workings of his setup, Teller realized that from his studies, he was quite familiar with the biological processes that happened inside the pit, making all these fortunes possible. At that moment, he knew he had found his calling. "That's what I want to do with my life now," he remembers thinking. "I want to build systems like this and make people's lives better."

He returned home, finished his degree, and pursued his goal with a vengeance.

He set up digester systems for the desert Bedouins, who shepherded goats and sheep, cooked on open fires inside their tents, and had some of the highest asthma rates among children anywhere—90 percent. He traveled through Oaxaca in Mexico and built digesters together with the local villagers. But after some time, he began to realize that these systems weren't always easy for people to implement and maintain.

The idea of a biogas manure pit may seem simple. You dig a hole in the ground two or three meters in diameter, you line it with bricks and concrete, you build a domed roof under which gas can accumulate, stick in a pipe, and you have your digester. But its actual construction and maintenance are more complicated. These bulky systems require a substantial investment of physical labor and money to build and operate. Depending on the availability of labor and funds, these big domes, with capacities of about 10 cubic meters, may take a few weeks to build—or even a few months. And they aren't necessarily the most hygienic of structures. If they are not tightly covered—and in many places, their tops consist of hay or wood planks—they may stink like enormous outhouses and serve as mosquito breeding grounds.

Moreover, often these concrete pits or domes don't last. After a few years, the ground may shift, causing cracks. Floods and frosts cause more structural damage, and the systems begin to leak, creating more odor and sometimes even fire hazards, because methane, of course, is combustible. Often the owners, who may have had enough money to pay for the initial installation, but not enough for maintenance, don't have the means to keep them up—and the digesters fall apart. When built right, they can last for 30 years, but if the concrete wasn't mixed right, cracks may start as soon as the next season. There are 40–50 million biogas domes around the world, in China, India, Africa, and Mexico, built as part of various programs to bring fuel to rural settlements. But it's estimated that about half of them are not working and need repair.

Teller realized that if biodigesters were to be widely used, they had to be less labor-intensive, easy to set up, and low-maintenance. He had seen how quickly villagers embraced solar energy projects—because solar panels came prefabricated, were easy to assemble, and required no hard physical work to install. In many cases, the locals were willing to pay for an easy-to-use product, but digesters weren't as easy as solar panels. Teller began to wonder if he could build an out-of-the-box digester that people could buy on the internet with a few clicks. Compared with a cement hole in the ground, it would be a radical idea.

One day, while visiting his 100-year-old grandmother, Teller explained to her what he did for a living. He described his systems, the complex reactions that happened inside the digesters, and all the benefits they were producing. Grandmother listened. And then she asked, "Do you have this system in your house?"

Teller felt humbled. He didn't have a digester at home. How could he build one? He lived in a modern village where digging a hole and filling it with waste was out of the question. Nor did he have animals that would produce enough manure. And no apartment buildings or urban sanitation authorities were willing to set up waste pits.

"My grandma told me to walk my talk," he says—but there was no way he could implement this idea in the so-called built environment.

Or was there?

Enter HomeBiogas

I'm sitting on a biotoilet. It has a comfortable seat, made of white plastic, that looks and feels pretty much like the seat of a Western toilet. It's also private—it's inside a nice log cabin that smells of cut wood and the plants that grow around it, rather than the typical outhouse odors. There's also a sink, the water from which is used to flush the toilet—although the traditional word "flush" doesn't quite apply, because it works with a manual pump. On the wall hangs a colorful poster explaining how a biotoilet works.

"Do your business as usual," states the first picture, portraying a toilet-perched millennial, pants down and smartphone in hand, tapping away. "It's OK to use toilet paper," says the next drawing, featuring a smiley poo pile waving the said paper roll. And there is indeed a paper roll hanging nearby. There's also a logbook pinned to the wooden wall, in which every toilet visitor must record his or her activities—how often they use it and how much they produce—small, medium, or large. These data are needed to calculate the poop-to-biogas ratio. After "your business as usual" is done, a simple yet clever system of pipes, pumps, and levers propels the waste from the toilet bowl into a reservoir where bacteria break it down, converting it to fertilizer and biogas. But while the biological processes are similar to those in Loowatt's biodigester, the implementation, and especially the packaging, is quite different. Simply put, it's personalized—built to accommodate an individual family, not an entire neighborhood or village. After all, the big necessity is your personal business, isn't it?

The biotoilet I am using belongs to HomeBiogas, headquartered in Beit Yanai, a cooperative agricultural community facing the Mediterranean Sea, about 40 minutes from Tel Aviv. Founded by Teller and his friend Oshek Efrati, HomeBiogas makes personal digesters the size of a big chair or a small loveseat. If I stick my head out of the toilet cabin, I'll see the digester standing right next to it, on the side, a green-and-black structure made of PVC—light and durable polyvinyl plastic. It has a tall, neck-like pipe for loading refuse, round sides puffed out with all the biogas goodness it contains, and a sand-filled bag on its top to provide pressure for steady gas flow. With the humpy sandbag, it looks like a camel resting on the ground. Teller says it's a fitting description. Camels store energy in their humps, and so do personal digesters. And if you feed them right, they can last for a long time.

HomeBiogas is Teller's answer to the need for a system that's easy to use and maintain. Originally, however, these personal digesters weren't meant for toilets. Teller first built a digester for food leftovers. "Food waste is a much safer topic," he says. People are scared of poop, but no one is afraid of stale bread or banana peels. Yet these scraps, thrown in the garbage and landfilled in many cities, are excellent biogas makers. "A food waste digester is even more efficient at biogas production, because it uses food that hasn't been eaten, so no nutrients have been extracted from it," Teller explains. "The nutrients go to feed the microbes. That makes microbes happy and they produce more biogas."

Teller and Efrati wanted these personal digesters to be the exact opposite of the unwieldy, high-maintenance in-ground manure pits, which worked for farms, but not individual families, not at the household level. They envisioned these systems as small, light, portable, inexpensive, and easy to ship and install. The digesters would have to be fully sealed, non-smelly, and made from modern materials that would last for years without repairs. "We wanted to build an IKEA furniture type of biogas system," Efrati says. "You order it online, you pop it out of the box, you follow instructions—and it works. Nothing like this existed yet, so we saw this gap and we knew we could fill it."

The idea was that any family, whether rural, urban, desert, off the grid, or in any other setting anywhere in the world, could buy this system and toss in kitchen scraps, plant leftovers, and other biomass—and never breathe smoke again. They could also toss in some dog droppings, chicken poop, or cow manure, but generally HomeBiogas systems were meant to process food leftovers, rather than excrement.

The first model, which the two partners built in 2014 with some grant money from the Israeli Ministry of the Environment, was bulky and cost $3,000. It was the size of a campground tent and looked like one, too. Teller and Efrati installed these tent digesters in Bedouin camps, and the systems began to chug along, consuming food scraps and puffing out biogas. "They're still in use," Efrati says—but they can process only small amounts of leftovers and need a lot of maintenance.

A year later, HomeBiogas sent a few newer models to Palestinian camps as part of the Partnership for Peace Program. As part of the project, Israeli and Palestinian students worked together to set up these systems for nomads, Bedouins, and other people in need. These improved digesters were smaller and cost $2,000 to build, but that was still expensive. The price of the third-generation models dropped to $1,000, and some were shipped to Gaza on behalf of the United Nations and the Red Cross. They still looked a bit like campground tents, albeit smaller ones. The camel-like systems are the latest and greatest iteration. They cost about $700, can hold 700 liters of gas, and are so compact that they can fit into my minuscule New York kitchen—although they are meant to operate outdoors.

While Teller and Efrati were busy perfecting their systems, their customers were busy using them. And then something unexpected happened. As the models grew smaller, slicker, and cheaper, Home-

Biogas customers also wanted them to be more versatile. They began asking Teller and Efrati to make the digesters process not only food waste, but toilet waste, too. They wanted to connect the camels to toilets.

The digesters really didn't care what type of biomass they were fed. The tricky part was figuring out how to flush human waste into them with very little water so that the miniature camels wouldn't overflow. "So many customers asked us for that—and at first we were really not sure about it," Efrati says. But then Teller moved to an off-the-grid community that didn't flush. Instead, the residents used composting toilets, mixing waste with other biomass and letting nature take its course. This setup wasn't immune to odors. In summers, it stunk. "Now Teller really needed a toilet," Efrati chuckles—so he figured out how to attach the camel system to one.

To solve the flushing problem, Teller equipped his toilet with a manual pump, which needs one-sixth of the water used to flush a standard toilet, but propels the waste and paper into the digester nearby. On average, the waste takes about 60 days to be decomposed by the bacteria inside, and it comes out in the form of liquid fertilizer, slowly accumulating in a container—a bucket or a watering spout. It reminds me of the brown goop children mix up in their play buckets when they make patty cakes and other make-believe food. And unlike the buckets my grandfather carried around, it smells like soil, not sewage. At the same time, biogas accumulates, providing fuel.

Efrati and Teller maintain that the liquid fertilizer is completely safe to use in vegetable gardens—and in fact, they use it to fertilize their own vegetable patches at home and office. But if users are concerned about germs, they can add a chlorine tablet at the end nozzle for disinfection. As the fertilizer passes through the tablet, the chlorine kills the pathogens, but while the liquid sits in the bucket, the chlorine evaporates, so when the resulting fertilizer is poured over plants, it no longer contains the toxic chemical.

The biotoilet became a research project. "It may sound funny, but this is real science," Teller says, showing the logbook on the wall as proof. If the camels are to work with toilets on a commercial scale, HomeBiogas needs to have all the numbers, specs, and limits. Right

FIGURE 4. Sewed in Beit Yanai, Israel, this HomeBiogas personal digester, designed by Yair Teller and Oshek Efrati, costs about $700 and can be shipped nearly anywhere in a compact box. Similar models can be attached to biotoilets to produce fertilizer and biogas. CREDIT: LINA ZELDOVICH

now, it's still in the information-gathering stage, says Lillian Childress, the company's research and development scientist, who came from the United States to work here, and who keeps track of the activities recorded in the logbook. There's a regular, flushing toilet in the office, Lillian tells me, but most HomeBiogas employees prefer this one. "The indoor one is inside a small room with no windows and poor ventilation, so it gets really stuffy in there sometimes," she says. "Here the air is always fresh." She was right. This toilet smells of forest rather than air fresheners. It's definitely more pleasant.

Using the manual flush, however, proves tricky. The poster on the wall explains how to use the lever, but I find myself in a battle with the flushing pump. After a few awkward moves, I eventually win the battle, wash my hands (which replenishes the pump's flushing water supply via a small pipe leading from the sink to the toilet), record my activity in the logbook, and head out to the HomeBiogas office.

We Fed You, Now Please Feed Our Camels!

The HomeBiogas "office" is nothing like a typical office. It's actually more like a campsite with a digester factory, a packing shop, a conference room, and a well-stocked outdoor kitchen, which of course cooks, boils, and grills using homemade methane. Like a little green-and-black herd, camel-like digesters are everywhere—set up for experiments or data gathering. Inside the factory space, three women in traditional Arabic scarves are stitching together digesters using industrial sewing machines. Yes, the camels aren't *made*, but rather *sewn*, then packed into compact boxes and shipped around the world. That's what makes them so cheap and quick to produce. Compared with a concrete-lined hole, they're a game changer.

HomeBiogas is hosting a conference, and the participants make up one of the most international groups of people I've ever seen. There are three women here from the Philippines, where the government is interested in distributing waste recycling technologies to the masses. There are two big, tall men from Ghana, who talk about the sanitation crisis in their country and their hope of turning it into an energy stream. A businessman from the Ivory Coast who runs a palm oil company wants the little camels to convert his rotting biomass leftovers into biogas. A Western doctor who believes in philanthropy wants to bring the camels to Nepal, which currently depends on India for its gas supply—but India periodically shuts off the flow, forcing villagers to cut down their forests. An appliance reseller from Brazil thinks this system will be very popular at home because it can save people money on energy bills. And there's even a computational geo-scientist from the Los Alamos National Laboratory in New Mexico, whose work on the melting Arctic permafrost and the subsequent greenhouse gas emissions from the bared earth led him to investigate whether anaerobic digestion could be put to common use. "I was thinking of building a digester myself," says Elchin Jafarov, "but then I found HomeBiogas and decided to try it."

What unites these people is that they all have a dream. They want to solve a waste problem. And they're willing to travel halfway across the globe to a "boot camp" where they will learn how to set up and run these digesters so they can adopt this technology back home.

Teller and Efrati say, first things first. "You traveled from far away. Let's have lunch and get to work afterwards." They urge us to dig into stuffed squashes, okra, stuffed grape leaves, vegetable rice, and an incredibly savory hummus. Lunch makes a great demo. It proves that biogas burners can cook just as well as the latest high-power stoves. As we wrap up our lunch, the duo encourages us to use the biotoilet, too. We fed you, so now please feed our camels, they joke, reminding me of the savvy sixteenth-century Japanese farmers who asked their guests to be generous and contribute. It's part of the experience, they add. Please make the circular economy work. "See that?" Teller pats a fat, puffed-up camel on the side of the biotoilet. "All this gas is coming from shit!"

With that, the conference starts.

HomeBiogas Lands Close to Home

About two months after Jafarov came home from the biodigester boot camp on the shores of the Mediterranean, his HomeBiogas system arrived in a small box, along with a 43-page instruction manual. Jafarov extracted the familiar green-and-black PVC camel from the box, set it up in his sunlit New Mexico backyard, and filled it with water.

Jump-starting the camel required manure. In theory, HomeBiogas packages would include a microbe starter kit to save squeamish biogas enthusiasts from handling a messy manure sample, but this early shipment did not. Once the system was operational, it could process any kind of organic waste, but the original set of microbes could come only from manure, not from food scraps.

So Jafarov drove to a nearby stable, picked up about 25 pounds of dry horse manure, brought it home, mixed it with water, and loaded it into his digester. As the manual predicted, the microbial zoo took about three weeks to multiply, and then the camel system swelled up with gas, pumping it to a burner. The entire family came to watch as if this was some kind of magic. "There was so much excitement when the gas came out," Jafarov recalls. "It's a very cool feeling when you turn on the gas and it works. We've been cooking with it all summer."

The camel became Jafarov's pet project, albeit with far-reaching goals. When winter came and night temperatures started dropping, which slows down bacterial activity, he moved the camel into a small greenhouse he built himself. He began equipping it with various sensors to measure its temperature and other vital signs. He wanted to see how viable the system would be year-round, and to estimate how much it could save a family in energy costs. For the United States, where energy is cheap, it wouldn't be much, but for other countries—for example, Brazil or Mexico—it could be a significant amount. The reseller in Brazil has been advertising these camels for almost double the HomeBiogas price—$1,300—and they are selling.

In the Western world, HomeBiogas may be a boon for those who want organic fertilizer for their gardens. "At my local farmers' market, eight ounces of organic fertilizer cost $8," Jafarov notes. It's very expensive. At these rates, the HomeBiogas system would pay for itself within days—it fills a 5-gallon bucket within a week. "I don't have a garden myself, so I've been giving the fertilizer to my friends who do, hoping that they may get interested in the idea."

Jafarov hopes that he will eventually be able to get people interested enough that he can start selling HomeBiogas systems himself. He has bought eight more systems that he hopes to sell eventually, at his cost price of $650. "I am not looking to make money, but to introduce people to the concept," he explains—because it's an environmentally beneficial way to process waste. So far, he has advertised HomeBiogas systems on his BiotechSavvy website and has verified that they do not pose any fire hazards. There are no regulations that prohibit their use, which makes sense, he says. At any given time, the systems contain such a small amount of gas, kept under such low pressure, that even in the case of leakage, the camels wouldn't cause any accidents. They are less dangerous than the LPG tanks people use for grilling, which contain liquefied gas under pressure.

Meanwhile, back at Beit Yanai, Teller and Efrati are gearing up to launch the biotoilet-digester duo. These systems, too, will be light-weight, user-friendly, and come in small packages. But will they work in the built environment?

Getting urban dwellers interested in converting their kitchen scraps to energy may not be hard. But convincing them to process their own waste might be harder. It would require installing a separate, probably outdoor, toilet, which would be difficult to do in urban settings. Even a chlorine tablet at the end nozzle may not convince the city sanitation authority that this technology is safe.

Could there be solutions for the city dweller on an industrial level?

Turns out, yes. In fact, your local sewage treatment plant may already be using some of them. When was the last time you stopped by your local waste processing facility?

It's time to pay a visit.

CHAPTER 10

MADE IN NEW YORK

"Welcome to the Valentine's Day tour of our Newtown Creek Waste-water Treatment Plant! My name is Ali and I am the plant superintendent of this facility."

That's how Zainool Ali, a smiley guy in a bright red shirt, greets the audience of about 200 New Yorkers whose idea of a romantic Saturday experience is visiting their local sewage treatment spot. "We do this every year," Ali adds as he welcomes us into the room and urges us to take our seats so he can start his presentation. "Those of you who came today as couples, this is the best place to be on Valentine's Day," he quips, cracking up the audience. "And for those of you who came as singles, I should say that in the past some people came as singles and left as couples. We hope you will too."

Would New Yorkers really look for mates at a wastewater treatment facility? It's hard to tell, but one thing is true: the tours are immensely popular, and the tickets sell out within hours. The plant hosts these programs only three times a year—on Valentine's Day, Earth Day, and Halloween—and the Valentine's Day occasion boasts the most tours. To sweeten up the subject, a young lady at the door gifts everyone with a little Hershey's Kiss.

New York has 14 sewage treatment plants spread out through the city, which together clean up and process 1.8 billion gallons of waste-water a day. Newtown Creek, the largest of them all, contributes about 350 million gallons daily, but can chug double that in heavy rains. Located in the Greenpoint neighborhood of Brooklyn, it's the most advanced, modern, state-of-the-art wastewater treatment facility in the city. It took 10 years and $5 billion to build—or rather, to rebuild the old, outdated plant that preceded it.

Known for its futuristic, gleaming, silvery egg-shaped silos, it is also the city's most beautiful and architecturally impressive sewage processing facility. The eggs, visible from many vantage points, are now favorite Brooklyn landmarks. The glass-enclosed walkway that connects the eggs' shimmering pinnacles offers visitors a bird's-eye view of the surrounding neighborhood from 10 stories up. The locals point to the eggs from their office buildings. The tourists inquire about them on the streets. Depending on a plane's approach to the JFK Airport, its passengers can sometimes spot them from the air, too. At night, they are lit up with different colors, as if to challenge the Empire State Building from the far land of Brooklyn.

These eight head-turning eggs aren't just for show. They are biodigesters that seamlessly convert New York's sewage into biogas. At 145 feet high and 80 feet in diameter, each egg holds about 3 million gallons of sludge, which it transforms into gas at a rate of 2 million cubic feet per day. That's part of the reason why some airplane passengers stare down at the eggs as they land: they are sanitation engineers coming from all over the world to visit the Newtown Creek plant and implement similar setups back home, Ali tells us. But the technology isn't cheap. Each egg cost $25 million to build and three months to assemble.

No one I ask seems to know how the tour got its Valentine's Day start, but one hypothesis is that it had something to do with the eggs resembling the shape of Hershey's Kisses. Or it may have launched spontaneously just because people wanted to see this impressive facility and stare down at the Brooklyn streets from tanks full of brewing dung. There are several legends floating around about the tour's beginnings, but its exact origin remains a mystery.

The plant's inner workings, on the other hand, are well known and documented, and Ali is here to tell us all about them. He promises to walk us through the journey of New York City sewage right in this room, figuratively and—almost—literally. His presentation table displays about a dozen bottles holding today's sludge and wastewater samples taken at different points on that journey—from dreadful-looking muck to clean, drinkable H_2O. Come here, grab your Hershey's Kiss, and ponder where the results of your digestion will go afterward, Ali banters. Yes, you guessed it—into one of those gleaming eggs.

On New York's Sewage Journey

Ali flips off the light switch, and the room goes dim. A projector turns on, and the screen on the wall glows with a map of New York City's sewage plants. "Depending on where you live, you can see who is lucky to receive your waste," he tells us, while the audience chuckles along.

Ali begins his talk with a bit of history. Originally laid in the flatlands of Brooklyn, the New York City sewage system was one of the first in the United States, built after the cholera epidemics of the nineteenth century. Just as the London engineers decided to build a combined water and sewage system, so did their New York counterparts. Not surprisingly, the New York system experienced similar problems, which are still pressing today. During heavy downpours, which are common in New York, the sewage system overflows, spilling untreated wastewater into the sea—and then the beaches display warnings to keep bathers away for a few days.

In recent years, the city has built many additional water catchments to store these temporary surpluses of rainwater, which has helped. But, Ali tells us, treating both the wastewater and the rainwater is expensive and not really necessary. "We have to treat both because our infrastructure is that old," he explains. "In a perfect world we wouldn't treat both." So the hope is that one day we will fix the mistake made by our ancestors by building two separate systems—one for sewage and one for rainwater, as was originally envisioned by Sir Edwin Chadwick and other British "sewage separatists."

When the pipes were first laid in the nineteenth century, there was no concept of sewage treatment—the outflow was simply dumped into the sea. But then New York Harbor began to stink—just like the Thames. As the city's population grew, the outflow also began to affect beaches and oyster beds. Studies found some dead zones in the ocean—places where the water is so devoid of oxygen that it has no marine life left. These observations prompted the start of the treatment era around the 1890s. The Coney Island plant, equipped with treatment faculties to protect the bathers frequenting the area's beaches, was one of the early adopters of treatment technology, says Dimitrios Katehis, director for regulatory compliance, strategy, and innovation at the New York City Department of Environmental Protection. But ultimately, the ocean remained the sewage's final destination for a nearly a century—until the US Clean Water Act of 1972 forced treatment plants to clean up their act. Sewage treatment underwent a major upgrade. Today it involves a number of different steps, which are represented by the curious array of bottles on Ali's table.

Ali picks up the first bottle, labeled "Raw Unscreened Sewage," in which a bunch of unidentifiable scraps float in a grayish liquid. On a closer look, one can discern a torn piece of jeans and a few glops of white fluff, which Ali says are probably baby wipes. The latter is a sanitation engineer's nightmare. The packages tell you that baby wipes are flushable. But that doesn't mean they are biodegradable, he explains, so instead of dissolving like toilet paper, they whirl up in balls, clogging the system. Add them to the plastic containers, glass bottles, and other street trash that gets washed into the pipes during rains, and you get a pretty wild array of dregs arriving at the plant's gate on a regular basis. To remove all this trash, the sewage is "screened." Underneath the plant, the sewage river flows through a gate or a rake, which sifts out the debris.

Ali lets us stare at the bottle in awe for a few seconds. And if this looks bad to you, he tells us, you'll be amazed to know what periodically shows up at those screeners, especially during heavy rains. A couple of months ago, the gates stopped a strange object that looked like a huge ball of yarn. It was so puzzling that it roused Ali's curiosity—and he began to unwind the yarn. "I wanted to know what it was, so I sat there with a rake and pulled at it," he recalls. After a while, two handlebars peeked through. Ali did more pulling—and out came the rest of the bicycle. "We never figured out how it made it into the sewers, but it's just one of the million things that turn up," he says.

Another recently common type of sewage debris, particularly at the Newtown Creek location, is construction materials, because of the recent construction boom in Greenpoint. When those materials aren't properly secured, heavy downpours can carry them a long way, and they can easily pass through the sewer pipes, which are big enough to fit buses and trucks. "We used to get a lot of plywood, beams, and lumber after heavy rains," Ali says—all of which can clog the system and ruin the processing machinery. "We had to man our machinery. As soon as we saw something suspicious, we had to stop the processing equipment."

Once the big, hazardous pieces are cleared out, the next step is to remove the "grit"—a substance presented in bottle number two. Grit is essentially a variety of smaller particles that are still too big for further processing and must be filtered out.

More bottles follow, depicting several progressive stages of the sewage life cycle. The sewage passes through aeration tanks, where aerobic microorganisms break down organic compounds and nutrients. Then it's disinfected with chloride—a stronger, tougher relative of common household bleach—and treated with other chemicals. And finally, the effluent is treated to remove excess nitrogen—that ticking time bomb that slowly ruins waterways and marshes.

New York City's sewage plants are quite conscientious about preventing nitrogen from spilling back into nature's waters, explains Katehis. And they use a surprisingly familiar and simple compound to extract that nitrogen from the wastewater: glycerol. Historically a common ingredient used in manufacturing pharmaceuticals, glycerol has become the sanitation industry's new darling. Glycerol contains carbon (its chemical formula is $C_3H_8O_3$), which certain types of wastewater bacteria can feed on. As those bacteria eat up the carbon, they also pull nitrogen out of the water and transform it into gas, releasing it back into the atmosphere, where it naturally occurs. This bacteria-based approach is called biological nutrient removal, or BNR. At the end of the process, clean, denitrified water flows into the East River and eventually to the ocean.

How clean is clean? Ali demonstrates. "Can you flip the lights on? I want to make sure I get the right bottle," he says to a colleague standing next to the switch. When the lights turn on, Ali picks up the last bottle, labeled "Final Effluent," unscrews the top—and takes a few long, indulgent sips. The audience releases a communal gasp. That's how clean, Ali says. But water is the easy part, he adds. The remaining sludge is where the fun really begins—and a couple of remaining bottles on the table will help him to demo that fun.

The problem with the remaining sludge is that it's, well, sludgy. It's neither liquid nor solid, and therefore it's hard to process, pump, or drain. So treatment plants commonly add a coagulant: an acrylamide polymer that thickens the sludge into a more manageable form, which can be pumped into a centrifuge to spin out the remaining water.

That simple step has some unexpected financial and international consequences, Katehis reveals. Over the years, the production of coagulant compounds has shifted to China. Now all these materials are imported and are therefore subject to trade tariffs. Whenever a trade war breaks out and the commodities market is shaken, coagulant prices can go up overnight, spiking sewage processing costs. In the morning, unsuspecting New Yorkers will shuffle into their bathrooms just the same, but their every flush will be more expensive than it was the night before.

"These coagulants are critical for us because we will not be able to convert liquids to solids without them—and we need to make things viscous so we can drain the water out," Katehis explains. "Every time there's a trade tiff between the US and China we have prices spiking up and vendors not being able to deliver, or vendors dealing with tariffs. It's a new set of challenges for us. It creates a lot of market volatility which we cannot afford—because people flush toilets anyway."

Ameliorating this problem is far from easy because of its scope. "We are the number one consumer of that polymer in the Northeast," Katehis says. "We try to minimize our exposure to the commodity markets by having either multiple venues or alternative processes, or all of the above. Part of our R&D effort is to develop technologies and also reduce costs and to limit our market exposure to have the peace of mind, so when the next trade issue with China happens, it doesn't affect us."

Regardless of how much it costs to thicken the sludge, it is destined for the silvery landmark eggs. Once loaded into the digesters, the sludge simmers there at 98°F for at least 15 days while various microorganisms eat through it. "Digesters are essentially a replica of what goes on in your stomach," Ali says. "It takes food and breaks it down. That's where the 98°F setting comes from—it is the temperature of the human body." And just as Loowatt's miniature biodigester plant uses biogas to heat water for sludge disinfection, Newtown Creek burns its own biogas to keep the tanks' bacterial community at this comfy body-like temperature.

The description of this microbial melee instantly piques my curiosity. So which bacteria live in there—the same kinds that normally inhabit the human gut? If so, the sludge wouldn't be dangerous to us, would it? But that just doesn't sound right, so there are probably other, pathogenic, species in it, yes?

Ali says that there's a gamut of bacteria inside the digesters, but he can't name the exact species or explain what they do, so I make a note to dig into it further. Besides, our lecture is coming to an end, and we are nearing the most exciting part of the tour: the digester field trip. We don hard hats and follow the Newtown Creek staff up to the glass observation deck linking the eight digester tops.

Our group shuffles out enthusiastically, despite the bitter cold—it's 17°F, and feels like 9°F in the wind. But the digester eggs are an irresistible draw in any weather. Although, we are told, it's a bit nicer up on the observation deck because the glass cuts the wind, and the vapor rising from the digester vents makes it a touch warmer than the surrounding frigid air.

On the way to the digesters, I begin to understand why the plant does so few tours despite the unceasing interest of the public. It's not easy to chaperone a couple of hundred curious New Yorkers about 10 stories above the ground, never mind the hazardous materials involved. Nor is it easy to herd them through the walkway's windy corridors while they snap their selfies atop Brooklyn. We have several chaperones with us, who make sure we take turns waiting for the slowly moving elevator and don't stumble over various pipes and valves.

The glass walkway proves to be a universe of its own, a sneak peek into the otherwise inaccessible wild and weird world of biodigesters. Sticking up from the floor, black air vents puff out warm vapor smelling of ammonia, sulfur, and other pungent odors—bad enough to make you hold your nose. The whitish whiffs rise up in the cold winter air like vestiges of a dormant volcano—or perhaps like the breath of some mystical beast sleeping underneath our feet. The latter is an apt comparison, because there is indeed a living creature beneath us—a community of trillions of anaerobic bacteria eating, inhaling, and exhaling in these tanks. I ask one of our chaperones if they know what bacteria are breathing out this cloudy mix of gases, but no one seems to know for sure. I am told that it's a complex mix of creatures doing some very complex things—and producing the biogas, which we collect from them.

I switch my attention to the digester tanks and their lids. Each tower top features a round well with a glass cover through which you can peek into the digester's dark insides. My first look, as I peer through the tightly closed glass circle enveloped in clouds of warm, whitish vapor, reveals nothing but an infinite black void. But as I continue to stare through the glass, I begin to discern some interesting movement within that dark liquid chasm. Underneath the glass, dark waves roll across the surface, like ocean swells. They are so mysterious and interesting that I hold my phone down to the glass and capture a few seconds of this particular digester's waves on video. That earns me a few curious looks from the other "Valentiners," and they begin to peer through the glass, too.

Where does that movement come from? I wonder. Bacteria can't make these waves, so there must be a mechanism of some sort keeping the liquid in motion—probably for better digestion. Mesmerized by the enigma, I keep watching the waves rise and fall, periodically obscured by the cloudy white whiffs. It's a fascinating thing to see digestion in progress and, equally, to ponder its inner workings. I'm peering into the brewing poo of about a million New Yorkers, so opaque that I barely discern a thing, a black abyss filled with trillions of bacterial cells—and no one can tell me exactly what they do and why. I'm staring into an infinite blackness of brewing, churning waste, safely sequestered from the humans who produced it—and I can't wait to dive into it. I need to get the bottom of that futuristic, shiny, and enigmatic pit—and learn all about the mysteries of biogas biology.

Dipping into History

The first thing I learn is that biogas production is a surprisingly old idea. People have known how to make and use biogas for centuries. The earliest historical mention of biogas use dates back to the tenth century BC, when the Assyrians used it to heat their baths.

The scientist credited with identifying methane's chemical formula in the nineteenth century was Amedeo Carlo Avogadro, an Italian scientist best known for his contribution to molecular theory, now called Avogadro's law. However, the first one to experiment with methane—in the form of marsh gas, which naturally forms in swamps—was another Italian physicist and chemist, Alessandro Volta, about 50 years earlier.

In the nineteenth century, Louis Pasteur generated biogas from horse manure—and proposed to use it in street lamps to keep Paris lit at night. At the time, the city was covered with manure from all the horses that lived there, so Pasteur's idea could kill two birds with one stone: clean up the city and put all the dung to good use.

The first anaerobic wastewater treatment plant that produced biogas was built in Germany over a century ago—in 1906. Germany embraced this new idea, and in 1920, built the first sewage plant that channeled its biogas into the public gas supply system. After that, biogas plants really took off, and the technology quickly spread from Europe to Asia. In China, biogas plants were first built by affluent families in the 1940s. India began building simple biogas plants for rural households in the 1950s. In Germany, biogas technology adoption was driven by the country's need for alternative energy sources in a war-torn economy. In India and China, there was strong government support for production of biogas as a cheap source of energy, especially in rural locales.

Cheap oil prices in the mid-twentieth century dampened the world's interest in biogas technology. However, following the oil crisis of the 1970s, it gained renewed interest. Right now, over a million biogas plants exist in India. In China's rural areas, more than 20 million people rely on biogas produced by over 5 million small digesters. We are simply rediscovering the wisdom of generations of engineers who had it all before.

Biogas Biology

Biogas technology appears simple, but the actual inner workings of a digester comprise a complex, multiphased, and tightly interconnected set of biological process involving numerous bacterial species, both harmless and dangerous. Even though some bacteria in the digester are similar to those found in human guts, overall, the microbial menagerie living in a typical digester can be quite pathogenic. Species isolated from anaerobic digesters include *Clostridium*, *Bacillus*, *Staphylococcus*, *Streptococcus*, *Pseudomonas*, *Actinomyces*, and *Escherichia coli*, all of which have pathogenic strains. But other beneficial microorganisms, including those that normally dwell in the human gut, have been found in digesters, too, including well-known probiotics such as *Lactobacillus*, which improves digestion, and *Bifidobacterium*, which actually fights *E. coli*.

Once I finally delve into the anaerobic digestion process, I understand why no one could explain it to me in just a few words. The molecular reactions stretching across pages of text are indeed so complex they can confuse a chemist unfamiliar with this particular topic.

As one paper explained, anaerobic digestion is a process in which organic matter is broken down by a consortium of microorganisms in the absence of oxygen, yielding simpler chemical compounds and biogas, which consists mainly of methane and carbon dioxide. But it's a chain of reactions, involving complex organic and inorganic chemistry, that scientific literature breaks down into three phases—or sometimes even four, because one of the phases can be split into two steps.

The first phase sludge goes through is hydrolysis. During this phase, large organic molecules dissolved in water—such as carbohydrates, cellulose, proteins, and fats—are decomposed by acid-producing bacteria into smaller compounds. Using specific hydrolytic enzymes, some of which are present in our own stomachs—like amylase and lipase—the acidogenic bacteria break down proteins into lipids and amino acids, and complex sugars into simple ones.

The next phase is called acidification, which some scientific literature splits into two steps: fermentation and acetogenesis. In this phase, the fermenting bacteria and the acid-producing bacteria work together. The fermenting species produce what the acid-forming ones need to break the organic compounds down further into amino acids, alcohols, and other compounds. As one might expect, these bacteria produce several types of acids—butyric, propionic, and acetic acid—all of which contain carbon in various forms. In the process, they also release carbon dioxide and hydrogen.

The hydrogen and the acids become food for methanogens. Scientifically speaking, these methanogens aren't bacteria. They are archaea, a different, older form of microorganisms that live only in anaerobic environments. The methanogens sit at the last phase of this microbial conveyor belt.

The methanogens like their food simple. They can't break down and metabolize complex organic materials, but they can synthesize methane from the hydrogen and simple carbon-containing compounds the acidogenic bacteria have produced. And although they are picky eaters, they are a rather varied species of at least several different strains with diverse appetites. They feed on hydrogen, acetic acid, carbon dioxide, and carbon monoxide—along with a few other carbon-containing chemicals—and convert them into methane gas, with the chemical formula CH_4.

The digester's bacterial communities are highly interdependent. Some of them produce compounds used by their neighbors. Some consume products that harm other species. For example, methanogens, without which biogas won't be produced, dislike oxygen—it's toxic for them. Even though digesters are anaerobic, some chemical reactions that happen in them release oxygen—and if too much oxygen builds up in the black liquid goo I was staring at, methanogens will die out instead of producing methane. The reason methanogens survive and thrive is that the acetogenic bacteria create a favorable environment for them. To produce acetic acid, acetogenic bacteria need oxygen, which they take from water as dissolved oxygen or wrestle out of other compounds, ensuring that there's no free-floating oxygen in the digester.

In a way, methanogens return the favor. The acetogenic bacteria don't do well in the hydrogen-saturated environment they produce. Their own waste product, hydrogen, would essentially choke them to death. But methanogens solve this problem—they gobble up hydrogen, incorporating it into methane and balancing out the system.

This overall digestive balance is very important. When either species releases too much of its own waste product, it starts to die out, producing stagnant zones in the digester. To avoid this problem, it's important that all the species be well mixed. That's why some digester literature recommends "agitating" the digester—mechanically mixing the contents. That's what those mysterious waves inside the eggs were about—they were made by some inner machinery that was mixing the microbes well.

In a well-mixed digester, these chemical reactions can essentially keep going forever—as they do in swamps and marshes, where Alessandro Volta found the marsh gas he experimented with. And that is one beauty of the digester reaction: once you get it started, it can essentially keep going forever, with Mother Nature taking her course.

(Over) Made in New York

One would think that New York's 9 million residents would be a boon for biogas production. If the Newtown Creek plant, which serves about 1 million people, generates 2 million cubic feet of biogas every day, then outfitting the other 13 wastewater treatment plants with digester eggs should produce an additional 26 million cubic feet daily.

FIGURE 5. These futuristic silvery egg-shaped silos at the Newtown Creek Waste-water Treatment Plant convert New York's sewage into biogas. At 145 feet high and 80 feet in diameter, each egg holds about 3 million gallons of sludge, converting it into 2 million cubic feet of gas every day. The plant conducts tours of the eggs on Valentine's Day, Earth Day, and Halloween. CREDIT: LINA ZELDOVICH

To put that in perspective, 1,000 cubic feet of natural gas is enough to run an average American home, including heat, hot water, and cooking, for four days. That means that the daily output of New York's 14 wastewater treatment plants could power 28,000 homes for four days, or 112,000 homes daily. So if we can figure out a way to channel this gas back into the grid, maybe we can even cut down on some fracking. And if Germany could master that technology in the 1920s, we should be able to do so, too.

The idea sounds great, but it's more complicated than it seems. In fact, Con Edison engineers are experimenting with routing the bio-gas into existing natural gas pipelines, but it's easier said than done. The biogas that comes out of the digesters is a mixture of methane and some "impurities"—for example, sulfur, which is a health haz-ard and an environmental contaminant. These impurities must be removed before the eggs' output can join the pipeline. Ali says that after testing, Con Edison found Newtown Creek's biogas to be "almost pipeline quality." The company built a pilot gas-cleaning plant—and it worked. Now Con Edison is planning to build a full-scale gas-cleaning plant next to the digester eggs—but like everything else done on a grand scale, it will take time.

Consequently, right now, Newtown Creek produces more biogas than it can use. It uses what it needs to keep the digesters warm and burns the rest, creating carbon dioxide.

Nonetheless, the New York City Department of Environmental Protection is investigating ways of making biogas production more efficient. Just as Loowatt adds food waste to its digester, New York is experimenting with adding some food scraps to the eggs. This would also help divert the city's organic food waste from landfills to the treatment plants, where it would be broken down and used to produce biogas and biosolids. But until Con Edison actually builds the gas-cleaning plant, the excess biogas won't be usable anyway.

The Newtown Creek biosolids also deserve better placement. Right now, the digested sludge from the eggs is pumped onto barges and shipped off to another plant, where it's further dewatered and turned into the final biosolid product, commonly called a "cake." During our Valentine's Day lecture, Ali showed us pictures of the sludge dock and the three motorized barges that shuttle that sludge. I can't help but think of the sixteenth-century Osaka vessels that so efficiently transported the city's output to the farmers in the country, despite the stink. Did we manage to re-create the wisdom of our ancestors and find a way to put that "cake" back into the fields?

Well, not quite.

"We landfill it," Ali told me. The cake essentially serves as a landfill cover, ensuring that the city garbage doesn't reek. So it joins food, trash, and other organic materials slowly decomposing into dirt. Some of that cake is also trucked to abandoned coal mines in Pennsylvania as part of the soil reclamation process. It's better than burning it or dumping it into the ocean, but it still doesn't truly fix the problem of redistribution of nutrients on the planet, because we still don't use these biosolids to grow food.

In the 1970s, there was a Bronx company that shaped sludge into a fertilizer product. And even as recently as the 1990s, New Yorker–made fertilizer was still a prized item, Katehis says. "Back in the day, in the 90s they used to put biosolids in rail cars, which would go all over the country and out west, and all the way to Colorado," he recalls. "The farmers there prized New York City biosolids. It improved the quality of drier soil because of the organics it contains." But then it became too expensive to ship biosolids so far. "And so we focused more on using them closer—in Pennsylvania, Ohio, New York State, and on using them for other purposes—like landfill covers."

It's a good thing the biosolids end up in the old mines, helping Mother Nature recover after what we have done to her. It's a good thing the cake works as a landfill cover, cutting the stink and perhaps helping all that trash to biodegrade better. But I still wish there was a better application for all this organic goodness, a better way to send back all these riches to the fields. I had actually seen the cake-carrying barges floating up and down the East River. So what's stopping us from being able to direct these barges to farm fields somewhere?

It's an issue similar to the biogas overproduction problem. The fertilizing potential of New York's 9 million residents is so immense that no market can readily consume what the city's population can dish out.

"We produce so much of the product because the system is so large that we would just oversaturate the market," Katehis says. "We must have more diverse strategies and find different users. We have to be careful not to inundate the market."

In other words, we are back to the same old problem of efficiently shipping night soil from cities to farms. Finding users for humanure close to a big city is problematic, and shipping it to faraway locales is expensive. So the solution becomes finding other suitable ways to dispose of it—such as landfills.

That's not to say that New York City doesn't invest in its sanitation infrastructure. Just in the last 10 years, the city spent over $10 billion upgrading its sewage treatment plants. The results are impressive: over 7,000 miles of sewer pipes, eight sludge dewatering facilities, and three sludge barges shuffling our shit to places where it can be reused—even if not in the most ideal ways.

But not every city out there has New York–sized budgets and New York–sized problems. What do the small rural places do? What do you do if you are a small town or rural municipality that doesn't have $10 billion, but also doesn't have to truck biosolids to farmers a hundred miles away? Can "rural-made" biosolids be put to good use? Is there a way to use them cheaply and sustainably—and in a better way than simply landfilling them?

This idea may be worth another trip.

CHAPTER 11

LYSTEK, THE HOME OF SEWAGE SMOOTHIES

There's a sewage river gurgling beneath my feet. Thousands of bouncy bubbles, large and small, float up to its brown surface and burst, just like the foamy fizz in a soft drink. Surprisingly, I can barely smell this river, but it's quite loud—the popping bubbles create more noise than a medium-sized waterfall. To be precise, this river is actually an aerated sewage tank with oxygen being pumped through it, part of an aerobic or open-air sewage treatment process. I look down at it from a metal scaffold above. With all that noise, I have to strain my ears to hear what John den Hoed, the plant's supervisor of wastewater services, acting as my tour guide, is telling me, and I have to shout to be heard.

Den Hoed manages a sewage treatment plant in Elora, which is part of the township of Centre Wellington, a small Canadian town about two hours west of Toronto. But despite its small size and simple setup, Elora's plant has a novel, cutting-edge technology that it uses to convert sewage sludge into organic fertilizer that not only conforms to the safety requirements of the Canadian Food Inspection Agency (CFIA), but also exceeds the safety requirements set by the US Environmental Protection Agency (EPA). The plant's technology, implemented by a Canadian company called Lystek, was developed by two local scientists and is now gaining popularity in both Canada and the United States.

As den Hoed talks, I stare at the aerated fecal matter that has recently arrived at the plant and wonder what it takes to render it harmless. That bubbling brown brew is full of life—bacterial life. There are ciliates and rotifers floating in it, some of which can be pathogenic and others that are not. There are nematodes—roundworms that humans are better off without. There are also spiral-shaped bacteria called spirilla, one of which can cause a type of rat-bite fever in humans (although that particular strain doesn't normally live in water). Apparently, tardigrades live there, too—the tiny but tough little organisms lovingly nicknamed water bears for their bear-like shape and known for their remarkable survival talents. Tardigrades aren't harmful to humans. It's nice to know that at least some microorganisms in our excrement are completely benign to us.

Outside of the truly dangerous pathogens such as *E. coli*, salmonella, cholera, or hookworms, these typical sewage dwellers—spirillum, ciliates, rotifers, nematodes, and tardigrades—provide us with important benefits. They actively eat through the sludge, helping transform it from pathogenic filth into fertilizer. They are the first step in that conversion process, so they are carefully maintained and shuffled between various tanks to make sure that they produce the next generation of sewage eaters before they die of old age. The water in the open-air tanks moves nonstop, so it doesn't freeze, even in winter, despite the bitter Canadian frosts.

Running this operation is as much an art as it is a science, den Hoed reveals. Sewage never stays the same—it changes every day, as people's meals, diets, and metabolites change. To me, the contents of the tank look like a big malodorous mess, but a seasoned eye and nose would spot some intricate nuances, he says. Some days the tanks smell slightly unusual. Sometimes the foam on the top looks different. The sewage-eating menagerie plays a role in these changes, so den Hoed and Anton Wasilka, the plant's lead operator, periodically drop in to see how the creatures are doing. They can view them on a big magnifying computer monitor hanging in the back office.

Once the sewage eaters have gobbled up their food, the sludge is pumped into a huge centrifuge that squeezes as much water out of it as it can. It thickens the sludge from about 2.5 percent solids to 17–18 percent solids—essentially to the consistency of wet clay. "It comes out like mud," den Hoed tells me. "When I was a kid, I played a lot by the river. You know that mud that you get on the bottom of your boots? That's what it feels like."

That's where Lystek's biofertilizer conversion magic comes in: the mud serves as feed for the Lystek system.

In the past, Elora's plant, as well as many others, would simply truck that mud, or the still-liquid sewage, someplace where it could be safely disposed of—out of sight and out of mind. These Western methods of sewage disposal don't really sound much better than those in Madagascar, Kenya, or other places. The one big difference is that the Western sewage is treated to kill the pathogens, whether thermally or chemically. But at the end, it is still *disposed of* rather than *used*. Until recently, this was the status quo.

But early in the twenty-first century, two curious scientists reasoned that wastewater treatment plants were foolishly wasting their waste. Their inspiration took root at the opposite end of the sewage cycle: in their own kitchens. This was the beginning of Lystek.

A Sewage Smoothie, Anyone? Solving the Biosolids Challenge

At some time early in the millennium, Ajay Singh and Owen Ward, two researchers at the University of Waterloo in Ontario, noticed something peculiar about the local traffic: there were a lot of really big trucks buzzing around the city of Waterloo, and they were doing so with remarkable regularly. What were they trucks carrying?

As it turned out, the trucks were hauling human waste. Or, in industrial terms, they were taking away and disposing of sewage. To Singh and Ward, this looked like an imperfect way to handle sludge. These gasoline-powered trucks were essentially transporting water—because water constitutes 98 percent of liquid waste. "You're basically using fossil fuels to move around water! It really does not make sense," Singh says. Some sewage treatment plants dewater the sludge, then purify the effluent and release it back into nature, but that still leaves them with the so-called biosolids, or in den Hoed's description, "black mud," which isn't any easier to deal with. "Biosolids" is a euphemism used to describe a black, malodorous, pathogen-laden muck that is too thick and difficult to pump through pipes and too hazardous to use as fertilizer. It's no surprise that no one wants it.

In North America, most forms of sewage, whether biosolids or liquids, are unwanted, so every municipality tries to get rid of them cheaply and quickly. In some cases, biosolids are incinerated or burned, spewing carbon dioxide—the greenhouse gas that contributes to global warming—into the atmosphere. In other areas, they are dried and tossed into landfills or abandoned mines. And in some places, they are channeled into so-called holding lagoons.

"Lagoon" is also a sewage industry euphemism. Holding lagoons have nothing in common with nature's beautiful inlets or coves. They are essentially giant human-made sewage containers, situated far away from towns or developments, to which liquid sludge is shipped and dumped in, and where it simply accumulates over time. The solid parts sink to the bottom, and the water layer floats on top, serving as a protective stratum that may cut down on the stink. In some cases, the contents of lagoons are used to fertilize fields, especially crops not intended for human consumption, such as hay, but often they just sit like monstrous pit latrines, thousands of times larger than those in Madagascar or Kenya. The liquid sewage from the trucks Singh and Ward saw in Waterloo was probably going into one of these pits.

"These lagoons are all over North America," says Kevin Litwiller, Lystek's director of marketing and communications. "In most municipalities it's just an accumulating liability that sits there until you can't put anymore material in it, and so they need to be dredged and cleaned up." When the content rises up to the brims, municipalities scurry for solutions. Some may actually have money put away because they know the day will come, Litwiller says, but many don't. Municipalities that have space dig new lagoons. Those that have neither space nor money still have to deal with them somehow, because their residents keep generating sewage no matter what. These lagoon cleanups can be expensive, just as it was for the Londoners to clean up the Thames in the nineteenth century. When one Canadian town had to clean up its lagoon recently, it cost close to a million dollars.

The whole system looked far from ideal to Singh and Ward. As part of their research, they had previously experimented with other types of sludge—such as the leftover by-products of oil refineries. They were wondering if they could come up with some way of converting the unwanted biosolids into something more desirable.

In many ways, biosolids are indeed desirable. They are mainly composed of lipids, proteins, and carbohydrates—the leftover nutrients we haven't extracted from our food. They also contain trillions of bacterial cells. The scientists wanted to figure out if they could somehow break down and homogenize that unruly glop while also eliminating all the germs.

Their first approach was using enzymes—chemicals that can naturally break down lipids, carbs, proteins, and bacterial cells, too. The idea did work, and some scientific papers were published, but it was too expensive to scale up to the industrial level. "Municipalities go for the cheaper options that fit their budgets," Singh says. "So you have to look at the economy of your solutions, too."

What would be a cheap and easy way to break down all the compounds in biosolids, plus bacterial cells, and blend them together into a manageable mix? The most economical solution came from the scientists' own kitchens: A blender! When a kitchen blender purees your berries, bananas, and yogurt into a breakfast smoothie, it does the homogenizing job quite well. So, they reasoned, if you puree some biosolids, add some alkali compounds to react with and break down bacterial cells, plus heat up the mix a bit to quicken the process, the whole homogenizing idea might just work.

That's exactly what Singh and Ward did. They bought a top-of-the-line kitchen blender, tossed in some preheated biosolids, mixed in some caustic soda as an alkali, and pushed the button. The blender whipped up the mix and yielded a nicely liquefied black, foamy mush. (It's important to note that when bacterial cells break down, they release some of the water stored within them, and that water helps with the homogenizing.)

The duo upgraded their kitchen blender to a high-tech industrial one. They used high-speed shearing, in which the blades rotate at 1,000 times per minute, replaced caustic soda with less expensive potassium hydroxide or lime, and heated the mix to about 70°C by pumping in low-pressure steam, essentially pasteurizing the goo. When their first high-tech blender produced the mud-like mush on an industrial scale, they named their creation LysteMize—from the concept of opti*mizing* the sewage treatment process.

The LysteMize method became the basis for the future Lystek's fertilizer production. "Plants need nitrogen, phosphorus, and potassium," Singh explains. "The sludge is naturally high in nitrogen and phosphorus, and we add the last element—potassium." And, when they added some other important plant nutrients, such as calcium, magnesium, and zinc, their resulting bio-fertilizer mix was christened LysteGro.

"We basically built a big milkshake maker," Singh chuckles. "It essentially makes a sewage smoothie every time it runs." And it's even pasteurized, too—just like milk in a milkshake should be.

The Big Milkshake Maker

At the Elora plant, Lystek's big milkshake maker is loaded straight from a centrifuge, which dewaters the original sludge just enough to make it pumpable—and the effluent goes back to nature. Once the blender makes the smoothie, the attached auger pumps propel it into holding tanks capable of storing a season's worth of it, where it awaits shipment to farms during the fall and spring fertilizing seasons.

Visiting the holding-tank building is not for the faint of heart. If the aeration tank smelled of fresh sewage, this building reeks like an old outhouse on the side of a road. The smell of ammonia, which hits me in the face the second I walk in, is dizzying. The tanks below me are full of revolting grayish goo, covered with grayish foam. That homogenized sewage smoothie is breathing, bubbling, and moving like one huge living organism. It's both repellent and mesmerizing.

I look because we have to solve the sewage problem one way or another—and companies like Lystek are trying to do just that. The slush down below is next season's fertilizer, all ready to be loaded into trucks and returned to the land to feed the next round of crops—as it ideally should be in nature. This is essentially a larger version of my grandfather's septic tank, or the Flemish in-ground sewage tubs, ready to feed the earth—only it's a twenty-first-century industrial version. It's still stinky, but it's a big move in the right direction.

When we leave the holding-tank building, den Hoed points at a truck loading station. with a big hose hanging down from the top. A truck drives in, the hose plugs into its tank, and the pumps do their job. "It takes about ten minutes to load a truck that holds 40 cubic meters of LysteGro," Singh says, explaining that this is, frankly, a re-markable speed. You can't do it with biosolids, because they are too solid, as the name suggests. They simply don't pump. It is Lystek's homogenizing process that makes them pumpable and loadable.

Once the trucks bring LysteGro to the farms, agricultural ma-chines called terragators plow through the fields and inject it four to five inches deep into the ground—closer to plant roots and far-ther away from the surface, where its smells may offend and worry humans. "We usually do two application seasons, spring and fall," den Hoed tells me. At the moment, it's been awhile since any Lys-teGro has been shipped off, so it's bubbling up at its high mark, den Hoed explains. All the waste diligently produced by Elora residents throughout winter is still sitting here in the tanks, waiting for spring.

I wonder if this is how the Flemish tanks looked when they opened them up in the spring and farmers showed up with their barges and horses. Were their tanks just as big? Elora has only about 7,000 resi-dents. Flanders had bigger cities than that, so it could be that some Flemish tanks held more. Then I wonder how the Flemish tanks smelled—did they reek of ammonia just as badly, or any less? If the Flemish collected their urine separately, their concoctions may have contained less ammonia and might have been easier on the nose. Ei-ther way, they had a working sanitation solution. Now, two centuries later, we are following in their footsteps on a grand industrial scale.

Lystek systems have already been installed in larger facilities. When sewage treatment plants want to stop dumping their bio-solids into landfills or mines, Lystek can act as an intermediary, ac-cepting that material as input and converting it to LysteGro.

That's what happens at the Southgate Organic Material Recovery Center—a large regional facility that processes the fecal output of about 100,000 people. Here, waste trucks bring and unload biosolids from multiple sewage treatment plants in the region—including some from Toronto. In simpler terms, the trucks dump piles of condensed shit on the concrete floor and drive away.

But the piles, which look just like that black river mud but reek unmistakably of an outhouse, don't sit there long. They are quickly loaded up into Lystek's high-tech blenders and converted to Lyste-Gro, which is then pumped into large, covered in-ground containers, where it's stored for the next application season.

And yet that's not all Lystek's sewage smoothie can do. It can also help fight eutrophication—that blight of nitrogen overload that ruins our waterways and marshes. And the magic is once again in the microbes.

Resetting the Nitrogen Time Bomb

Remember the dying Quashnet Pond, overfertilized by the nitrogen-rich effluent from Cape Cod's septic tanks? Municipal wastewater treatment plants have to battle the same problem: the excess of nitrogen and phosphorus in their effluent. If these two potent fertilizers aren't removed, they slowly overfertilize the accepting bodies of water, causing algal blooms and plant overgrowth. But removing nitrogen and phosphorus from water is tricky—there are no simple chemical reactions to capture them, and no sieve fine enough to filter out their tiny molecules.

Some modern treatment plants install so-called biological nutrient removal (BNR) systems, which use chemicals to induce specific bacteria to extract nitrogen and phosphorus from wastewater. Certain bacteria can feed on these elements, but they can do so only if they have carbon as a food source as well. If carbon is available, the bacteria accumulate phosphorus in their cells while converting nitrogen to a gaseous form and releasing it into the atmosphere. The problem is that wastewater is naturally low in carbon—because all the carbon, in the form of undigested organic compounds, is left in the biosolids. So, in order to make these reactions happen, wastewater treatment plants add in inorganic carbon-containing compounds, such as methanol (CH_3OH) or glycerol ($C_3H_8O_3$). But then the plants run into the same old problem of economics. Both compounds are expensive and add significant costs to BNR systems. Methanol is often the less expensive choice, but it's flammable and toxic to humans, which creates safety issues.

Lystek's smoothie presents neither of these problems. It happens to be high in organic carbon from undigested food, it's cheap because it's unwanted, it has no flammable compounds, and all of its germs have been killed. So when Lystek's carbon-rich smoothie is added to BNR systems, the microbes get loads of food, feel happy, eat a bunch—and in the process remove much of the unwanted nutrient content from the water. The nitrogen returns to the atmosphere, and the phosphorus accumulates in microbes' bodies, which can be collected when they die for fertilizing purposes.

Lystek's technology can fit easily into any existing sewage treatment plant. Similarly, Lystek's smoothie can be added to anaerobic digesters to increase biogas production.

When we come back to the Elora office, Wasilka gives me a close-up view of Lystek's smoothie. He brings up a little glass jar with a metal lid—one of those used for strawberry preserves. "It's a daily sample," he explains. "We check the product samples to make sure everything in it conforms to standards."

He takes off the lid and I peer into the jar. The black substance inside smells unmistakably of shit. But it looks like a shiny, creamy black butter, much more like shoe polish than processed poo. Judging by its look, it should have a very smooth, silky texture, pleasant to the touch. The smoothie piques my curiosity so much that I even consider putting my finger to its tiny black splutter on the jar's lid. It looks so small and shiny and benign. Can it be really that dangerous after it's been sheared, pasteurized, and treated with lime?

I almost do it, but recoil at the last moment. It isn't yet the black, composted dirt on my grandfather's farm, so it looks too scary to touch. Later, when we leave, I regret being afraid when I knew it was processed and pasteurized. I wish I had touched it. I have enough sanitizers on me to disinfect a cow. Then again, Wasilka told me that Lystek's smoothie is a very strong colorant—it can really stain your skin.

As I leave, I begin to wonder about something else. If this method works so well, why doesn't every city in North America install Lystek technology? Why don't our flagship cities like Washington, DC, or New York do something similar? Can you imagine the ecological difference it would make?

Turns out, our big cities are already making that difference, but they're doing it in baby steps relative to their size. It's not easy to overhaul a big city's sanitation infrastructure, so some are doing it slowly, plant by plant.

CHAPTER 12

HOW DC WATER MAKES BIOSOLIDS BLOOM

I'm standing on an observation platform atop 24 colossal pressure cookers. In front of me are fields of silvery pipes stretching over massive sewage aeration tanks. To my left is a building that holds a sludge-dewatering centrifuge. To my right are four tall, round concrete biodigester tanks, with capacities of 3.8 million gallons each. The pressure cookers below me hum steadily as they simmer. They are slowly cooking shit.

Welcome to DC Water's sewage treatment plant, which actually describes itself in a different way: as a resource recovery facility. There's a reason for that, says Christopher Peot, the plant's director of resource recovery, who is trying to change the mentality of the sanitation business. "There's no such thing as waste, only wasted resources," Peot says. "So we don't process waste here. We recover resources."

In addition to water, the plant extracts energy and a Grade A fertilizer product, sold in local stores, from the city's sludge. That's one thing those pressure cookers do—they heat and pressurize the sludge to render it harmless. They also make it easier to process and load into digesters. Scientifically, this process is known as thermal hydrolysis, and the huge, shiny metal cookers, each bearing the name "Cambi," thermo-hydrolyze the incoming poo, day in and day out.

The plant processes the contributions of about 2.2 million people who live in, work in, or visit our nation's capital and the area that surrounds it. There's input here from every house, every apartment building, every business, and every historical landmark—Tyson's Corner, the Smithsonian Institution, the Lincoln Memorial, and Capitol Hill. And the White House, of course. There's some presidential poop percolating in those aerating tanks, along with the input from the Senate, the House of Representatives, the Pentagon, and the protestors chanting in front of the White House lawn. "That's where all of us come together," jokes Bill Brower, a resource recovery engineer who is showing me around the plant. "And I'd like to thank all these people for their contributions."

One of the most ecologically savvy sewage treatment facilities in the country, the DC Water plant has an interesting history. The original facility, built the 1930s, had a digester that generated biogas, which was used to heat the plant. The digester worked into the 1980s. No one can quite pinpoint what happened then, but at some point the digester was no longer part of the operation. It was resurrected nearly two decades later, when the plant was due for an upgrade.

Peot likes to say that on his second day on the job he was part of the digester-planning meeting. The existing plant, which processed sewage using lime, was aging. It had to be either upgraded or replaced. Adding a digester to the existing setup would offer energy benefits. But Peot wanted more than that. He wanted the plant to generate a fertilizer that farmers and gardeners could safely put on their vegetables patches, a Class A biosolids product.

With a budget of $300 million, the plant's team sought bids for the technology it needed in North America. That didn't work out well. "There was only one bid, and it was 650 million dollars," Brower says. "Our board told us to look harder. So we asked, can we look beyond North America?" The board said yes, so the team looked across the pond—in Norway and England.

And they found Cambi.

Enter Thermal Hydrolysis

The original Cambi system didn't have anything to do with sewage or sanitation. First built in Norway, it was designed to break down paper industry refuse. The very name Cambi comes from the word "cambium," the part of a tree that fosters the growth of the cellular mass—essentially a layer of actively dividing cells that gives rise to other tissues.

When the paper and pulp industries process plant material, they generate thick, hard-to-manage cellulose waste. Cambi's creators, experienced in the science of thermodynamics and steam systems, devised their original steam explosion process to burst and break that tough cellulose mass into a more manageable form that could be easier to biodegrade. It soon became clear that the process could have other applications, and it was further developed into the Cambi Thermal Hydrolysis Process, or CambiTHP. The first Cambi system was installed at a sewage treatment plant in the Norwegian town of Hamar in the mid-1990s.

Shortly after, a wave of environmental regulations tightened sewage disposal rules across Europe, and Cambi piqued the interest of wastewater treatment companies in other countries. The United Kingdom was particularly interested in Cambi because, as it turned out, its sewage disposal methods hadn't made much progress since the days when Lord Bramwell's committee was cleaning up the Great Stink of London.

In the 1990s, British sanitation providers no longer had to scrape the foul-smelling sludge off the Thames banks, but they still disposed of sewage in the sea, says Bill Barber, Cambi's technical director in North America, who at the time worked for the United Utilities Water Company, which serviced the Liverpool and Manchester areas. The company had built a sewage pipeline all the way to the Liverpool docks. Through this pipe, it pumped the sewage into boats, which took it out to sea.

"These boats used to go into the ocean and open their hatches and the sludge would disappear into the sea," Barber recalls. "It was a common practice since the early 1900s, and it was finally outlawed in 1998." Ironically, the ban also wiped out some of the fish's favorite feeding spots, because the fish fed on the undigested nutrients in the sludge. "Somebody mapped all the places where they used to dump the sludge into the ocean and superimposed the spots where the fish were, and they found some correlations," Barber says. "The fish liked those spots because they were good sources of food."

When the sewage discharge ban went into effect, the sanitation companies needed a quick solution to comply with the regulations—and they turned to thermal hydrolysis. Thames Water was the first to try Cambi, and while the initial installation didn't work well (because it cut corners, Barber explains), when Cambi itself took over the management a few years later, it became a state-of-the-art operation. After that, the idea of thermal hydrolysis and biogas production took hold, and now about 60 percent of all UK sewage is processed that way, Barber says. Instead of pumping sewage into boats, his former employer now powers trucks with the biogas produced using Cambi and biodigestion. And it gives its fertilizer product to farmers, usually for free.

It was the Washington, DC, installation that brought Cambi its international acclaim. For the DC Water team, Cambi was a godsend—it was going to cost the company only $30 million—an order of magnitude less than it had originally budgeted. And after the technology was installed and proved to work well, other countries became interested, too. In 2014, the company signed contracts for 14 new plants, including in Spain, the Netherlands, and China plus an underground facility in South Korea. Beijing Drainage Group alone signed contracts for five large thermal hydrolysis plants to serve the city's population of over 20 million.

As Brower walks me down from the observation deck to a set of tanks labeled Heated Sludge, he explains their inner workings to me. The Cambi system consists of a preheating tank, called a pulper, and a series of upright cylindrical cookers, in which pumped sewage is heated to 300°F and pressurized to six atmospheres for a half hour. "That's six times more than what you are feeling now," Brower says. "That's how you kill all the pathogens."

Few living cells do well under these conditions. Most cook, implode, and die. Any that manage to survive undoubtedly give up the ghost when the cooked sludge is depressurized, causing the already stressed cell walls and cell clumps to burst and break apart. "That makes more food available for the microbes in the digester, and more biogas production," explains Brower. "That also changes the viscosity of the sludge." It makes the sludge easier to pump so that it fits into the digesters more compactly. As a result, DC Water built only four digesters, instead of eight. Building humongous concrete tanks is expensive, so cutting the number of digesters in half saved the utility about $200 million.

Before sludge goes into the Cambi system, it follows a pretty traditional route—the aeration tanks, the dewatering centrifuge, the effluent cleaning. Because the cleaned-up water is released into the Potomac River, which flows into Chesapeake Bay—an environmentally vulnerable body of water—the plant adds a robust denitrification step. The clean effluent becomes river water. And the dewatered biosolids are loaded into the Cambi system, cooked, and fed to the hungry microbes in the digester tanks.

Brower briefly recaps how digesters operate. "They work just like our stomachs. They take big, complex molecules like proteins, carbohydrates, and fats and break them down into smaller and smaller bits," he explains. "And just like bacteria in your stomach, bacteria in the digesters break down proteins into amino acids, which become food for the bacterial community further down the food chain, and so on."

Before the digesters could be used, they had to be inoculated with bacterial cultures. "We seeded our digesters with microbes from other digesters—actually from Alexandria across the river," Brower recalls. "We brought them in tanker trucks, put them in, and filled the rest with water—and then the population just takes care of itself." At the end of the process, methanogens produce methane, which is burned to spin the plant's electric turbines, generating 10 megawatts of power.

There's another great service these microbes perform: they dramatically reduce the amount of solid matter that needs to be dealt with. Without the digesters, the plant would generate 1,100 tons of biosolids, which even in their pathogen-free form still need to be trucked somewhere, using fossil fuels. The microbes reduce this amount by nearly two-thirds. "We went from producing 1,100 tons a day to 450 tons a day," Brower says. "The rest literally goes 'puff,' and not just puff but a puff that you can burn and create power," he points out. "Ten megawatts is enough to power like 8,000 homes in this area."

"But that still leaves you with 450 tons of digested sludge a day," I say. "Where does that go?"

"That becomes our premium product," Brower says. "That becomes BLOOM. Let me show you how it happens."

He hands me a hard hat with a blue-and-green DC Water logo, puts on his own, and heads into a concrete building.

Mining Black Gold

The building Brower takes me to is filled with conveyor belts. There are 16 of them altogether, and while not all of them are working, those that are on chug continuously, filling the air with a constant hum.

Officially called belt filter processors, these conveyor belts are designed to squeeze as much water out of the digested sludge as possible. Made from a permeable material, each belt essentially sieves the water out of the digested biosolids that are continuously deposited onto it. Think of it as a giant moving cheesecloth stretched out flat, only instead of white squishy curds, it's draining black soggy ones. And as the cheesecloth carries these glops along, the water drips down through it, so the soupy substance becomes less and less watery. It starts out like a muddy stream and drains to the consistency of wet forest soil.

At the conveyor's end, two roller presses flatten out this soil-like substance into shiny black sheets. The sheets quickly crumble into fragments resembling pieces of coal or graphite, only thinner and more uniform because they have been neatly pressed. The conveyor drops the fragments into a deep crater with an overhanging backhoe. It looks more like a coal-mining crater than a waste treatment facility, and it makes me think of a quarry. And then it hits me—of course it's a quarry. DC Water is quite literally mining black gold. All that treasure we have so foolishly been wasting in so many other places.

"It is very black," I tell Brower as I watch the fragments tumble down into the crater. "It's black gold!"

"I like that name, Black Gold," he chuckles. But this isn't the last step yet, he adds. A third-party company, named Homestead Gardens, takes this black gold, ages it, dries it up, and packages it in 25-pound bags labeled BLOOM. "We can make a more marketable material if we age it," Brower says—the moist product would stick to your boots and shovel, but the dry one is easier to handle; it's less clumpy and more powdery. He finds an open BLOOM bag and scoops up its contents to demonstrate.

I stare at the bag in awe. It takes up only a cubic foot of space, but it weighs 25 pounds—all that earthy goodness is heavy. But even more shocking is the fact that it bears no resemblance to shit. Unlike other processed sludge I've seen, like Loowatt's vermicompost or Lystek's LysteGro, BLOOM has no sewage quality left in it whatsoever. It looks just like everyday garden dirt.

Brower scoops up another handful, takes a sniff, and invites me to do the same. And because it looks so unlike shit, I realize I'm not afraid to touch it at all. I scoop up a handful of BLOOM—a dry, crumbly substance, which now looks a bit more brown than before, and its aroma travels up to my nostrils, triggering long-forgotten memories of the compost piles that my grandfather opened up in the spring. It smells of the harvest it can produce.

"What do you think?" Brower asks.

"It's musky and earthy," I tell him. "That's how forest soil smells if you dig a bit underneath the trees. I like how it smells. I like it a lot."

"You have a good nose," Brower says. "Musky and earthy is the number one descriptor we get. We have trained people who smell it and tell us what it smells like."

I dig into the bag with both hands and hold the small pile of BLOOM right to my nose. I draw a deep breath and close my eyes. The memories flood my brain fast, and I feel as if I am falling back through time. I see my grandfather again, alive and strong, carrying sludge buckets on his *koromyslo*. I see him standing on the porch, yelling at me to keep away from the pit. I see our dog, Bim, half mutt, half Siberian husky, barking excitedly. And I see our trees standing

FIGURE 6. Bill Brower, a resource recovery engineer at DC Water, looks at mounds of BLOOM, a Grade A fertilizer produced from the sewage generated by the residents of the American capital and its surroundings. He uses BLOOM in his own garden to grow squashes, tomatoes, eggplants, and other vegetables. CREDIT: LINA ZELDOVICH

tall, rustling in the wind, and strawberry patches swelling with red berries in summer, and bushels of golden apples in the fall.

"I grow everything with it, squashes, tomatoes, eggplants," I hear Brower say.

"How do they taste?" I ask. And then I turn away quickly because I'm afraid I'm going to cry.

"Everything grows great and tastes great," he says. "And I'm not the only one who thinks so. We've heard from a lot of people that they've got the best response they've ever seen from the plants. Particularly with leafy greens because that nitrogen boost does well with leafy plants. And the plants seem to have fewer diseases and fewer pests around—probably because BLOOM helps build healthy soils."

I can't help but think of Edwin Chadwick, who wrote how much tastier sewage-grown produce was—not only to humans, but even to cows, who clearly showed their preference for "sewaged" grass[1]—

and that makes me smile. Brower takes me to see the "growing facility," where the team tests the effects of different BLOOM mixes on plants. They mix BLOOM with various other organic materials—such as woodchips or mulch—and grow plants in the mixtures. Right now, the team is "in between growing seasons," Brower tells me, so there isn't much to show, but on the way out of little greenhouse structure, just before the end of my visit, we pass by a few piles of organic refuse that catch my attention. I recognize one pile as mulch, another as woodchips, but the third one looks both vaguely familiar and confusing. I can't tell what material the pile is made from. Then, on a closer look, I manage to discern letters on the pile's components.

"It's money," Brower tells me, seeing my confused look. "It is a pile of shredded money. When the Bureau of Engraving and Printing prints a sheet of, let's say, $50 bills and it is a little bit off, they shred the whole thing and burn it. We try to take the shredded paper instead of sending it to the incinerator and we mix some of it into our BLOOM mixes. And we do the same with other organic materials. So yes, by mixing poop and money, we can create something that has a lot of value. Something people would want to buy."

"Who buys BLOOM?" I ask. "Does it sell well? Or is the yuck factor still a problem?"

It's still a problem for some, Brower says, but it's a matter of educating consumers about the nature of the product and creating a market for it. Other companies have done it in the past. For example, Milwaukee Metropolitan Sewerage District has been producing Milorganite, a sludge-based fertilizer, continuously for over 90 years. Compared with New York City or Washington, DC, Milwaukee may have the advantage of having more farmland nearby, but homeowners buy Milorganite, too. So it's possible to create a market for a sewage-based fertilizer. "Homestead Gardens sells BLOOM bags for $15.99 in its stores," Brower says. "It's selling."

"What about farmers," I ask. "Do they buy BLOOM?"

"So far, we give it to farmers for free," Brower tells me. "And the reason we give it out for free is because we don't have our own storage capacities." There is only so much space to store BLOOM while waiting for farmers to request some, he explains. "We want to get our own storage facility and then we can change farmers, and I think they would buy it."

But there are also other niche markets—those that aren't put off by the long city-to-country haul or the yuck factor—such as landscape architects and construction companies. They buy good soil for horticultural use. When construction companies build new developments, they have to clear the site, cut the vegetation, and often strip the topsoil off. And once the buildings are built, they have to replace the soil and plant new trees, new grass, and new flower beds. BLOOM helps restore the damaged site—and it doesn't smell like manure, so tenants don't complain. "We can sell it because it's Class A fertilizer with very low odors," Brower explains—and there is interest. "Earlier today I got a phone call, and they wanted two loads tomorrow."

On the way back from DC Water, I hold my hand to my face and realize it still smells of BLOOM's pleasant springtime scent. Who would have thought that 40 years later, halfway across the world, I would witness my grandfather's process upgraded for the twenty-first century on an industrial scale, serving over 2 million people in my new country's capital. Or, more amazingly, that its end product would even smell the same.

I sit on the train sniffing my hands, oblivious to the curious glances of other passengers. If they had gardens, would they buy BLOOM? Some, maybe, others probably not. The yuck factor has been well cultivated in people—and for good reasons. So what else can we do with all that organic power within us? Can we turn it into other forms of black gold?

As it turns out, we can. And once again, all we have to do is look at what happens to organic matter in nature. Just as it seamlessly produces fertilizer and methane from our unwanted refuse, nature can produce other things. We simply have to mimic what nature does. It's not exactly simple, but it's not impossible.

CHAPTER 13

FROM BIOSOLIDS TO BIOFUELS

A couple of years ago, several Israeli researchers from Ben-Gurion University of the Negev set up a field toilet. Essentially a toilet seat stuck over a metal frame with a plastic-bag receptacle underneath, this kind of makeshift privy is an artifact of minimal convenience, typically used when scientists work in remote places where even portable johns and outhouses don't exist, or where the surrounding ecosystems are too fragile to process human waste. But in this case, the toilet itself was part of the research. That research was focused on the toilet's content and its potential to power human civilization. In simple terms, the scientists wanted to see if they could convert crap to coal.

Once the toilet was installed, several people used it daily, depositing their fecal matter in the bag. "We called them poop volunteers," says Amit Gross, chair of the Department of Environmental Hydrology and Microbiology at Ben-Gurion, whose team focuses on ways to transform waste to resources. At the end of each day, the collected matter was dried, then heated in autoclaves to kill the germs and make it safe for researchers to work with. Clad in face masks, gloves, and lab coats, the researchers pulverized the dried poo in a mechanical grinder—to reduce the yuck factor, Gross says. "Once it's pulverized it's easier to work with," he explains as he describes the challenges of their project. "This material can be quite repulsive for people. But in the pulverized form, you sort of forget what you are dealing with."

Once the matter was reduced to dark-brown powder, the real

experiments began. Researchers mixed the powder with water, doled it out into small batches, loaded them into nine 50-milliliter laboratory reactors, designed to operate at high temperature and pressure—and began cooking all that shit.

Cooks like to experiment with recipes to achieve the best results, and the Ben-Gurion researchers were no exception. They experimented with various cooking times and temperatures. They heated their batches to 180°C, 210°C, and 240°C. They let some of them simmer for 30 minutes, some for an hour, and others for two hours. Yet all the batches had one thing in common: they were cooked with water, but without oxygen.

Called hydrothermal carbonization, or HTC, this method produced hydrochar, a brown slurry of water and burnt—that is, carbonized—biomass particles. Unlike charcoal or biochar that's made by smoldering dry biomass, hydrochar is simmered in water. Because no heat is expended to evaporate water, this method takes less energy. "Removing water is energy expensive," Gross explains. "By leaving water in and using higher pressure and lower temperature, we use a lot less energy."

That hydrochar slurry has a double use. The particles can be separated into a powder with coal-like combustion properties, which can be fed into the industrial furnace of a coal-powered electricity-generating plant. And the liquid can be used as a safe organic fertilizer, since no pathogens can survive the cooking.

Cooking up human waste was a new project for Gross's team, but working with waste wasn't. His entire team focuses on waste repurposing ideas. "It's a different mentality," he explains. "We look at different wastes and try to think, what are they good for? Instead of thinking of how to get rid of it, we try to think how we can use it."

Gross's team's foray into the waste business started with poultry industry refuse. Globally, the world's poultry farms dish out between 625 and 938 million metric tons of litter a year, a number that will probably continue to climb, given the earth's human population growth. Poultry litter is high in carbon and nitrogen, the chemical elements needed for energy generation. And compared with cow manure, which may be spread out across pastures, poultry dung is amassed within enclosures. It was an attractive form of waste, so the team gathered droppings from a nearby farm, ground them into powder with a mortar and pestle, and cooked them up—dry, wet, at varied temperatures, and so on. The resulting hydrochar proved so similar to coal that it could be immediately sent to power plants, says Vivian Mau, another researcher on the project. It even came in the already powdered form, which is what many plants prefer, because they pulverize coal. "It's actually such a perfect fit that it can be fueling generators now," Mau says.

Hydrochar can also be pressed into briquettes for cooking and grilling, she adds. In fact, her team wants to do a poultry poop barbeque one day. They want to roast a bird on the hydrochar made from its own poop—and eat it afterward. Burning hydrochar doesn't produce any stink, but rather generates a pleasant aroma. "It smells of coffee," Mau says. "So we are really curious to see if the meat roasted with it will retain the coffee flavor."

According to the team's math, the world's poultry dung could replace approximately 10 percent of the coal used in electricity generation, reducing fossil fuel use. And the liquids could be channeled to agricultural fields—like the sewage pumped to the successful sewage farms of the nineteenth century, but fully pathogen-free.

With these promising results, human waste was naturally the next frontier for the team. It required an extra step to collect, with the use of the field toilet, but it proved worth the effort. Compared with bird droppings, human dung showed an even greater energy-generating potential. Unlike birds, which eat mainly plant-based foods, humans have very diverse diets, so their excrement contains more oils, which help combustion. "We eat a lot of fat, so that's probably why we end up with oily poop, and that's changing its properties," Gross says. "It's an interesting material to work with—not your typical biomass. We are currently studying its combustion properties, and we already see that the energy potential in the human material is higher than in the chicken manure."

Studying the waste's combustion properties means more than simply calculating its energy potential. When fuels burn, they release a variety of gases—not only CO_2, but in some cases, sulfur and other compounds. Some of these chemicals may be damaging to human health or cause environmental pollution. So before human waste can be approved as an energy source, it must pass various pollution tests. That is the team's next goal, Gross says. "We are currently evaluating exactly how human hydrochar burns, whether it would release any toxic gases and whether it would burn at a higher temperature," he explains. "We don't know it all yet, but we will very soon. At the moment, we just know that it potentially contains more energy per gram of excrement than other forms of hydrochar."

So far, the team has produced only tiny amounts of hydrochar, but scaling up shouldn't be hard, Gross says. A team of engineers can devise a working industrial solution to generate human-made coal and deliver it to power plants.

The reason hydrochar burns just like coal is that it's formed in a similar way. The conditions the Ben-Gurion team used to create it mimic, to a certain extent, the geological conditions within the earth that lead to the creation of fossil fuels. Fossil fuels, as their name suggests, are the fossilized remains of ancient organisms, dead and buried underneath many layers of other organic matter. Coal deposits began to form about 300 million years ago during the Carboniferous period, when the earth was covered with dense forests. When seas occasionally flooded the woodlands, trapping trees and other plants in swampy wetlands, the plants died and were buried under successive layers of vegetation, which turned into peat bogs. The ever-accumulating layers pressed the plant material deeper into the earth, where higher pressure and temperature, coupled with the lack of oxygen, slowly turned their organic molecules into the long-chain hydrocarbons that make up coal.

Oil, another type of fossil fuel, takes a similar, yet slightly different, set of conditions to form. Oil is the remains of marine organic matter—the microscopic plants and animals that lived in the ocean millions of years ago. When these bacteria, algae, and plankton died, they became carbon-rich deposits on the bottom of the ocean. Over millions of years, successive layers of such sediment deposits were formed on the ocean floor, and as they were buried ever deeper, heat and pressure rose even higher than inside the shallow peat bogs. These conditions transformed the dead biomass into hydro-carbons, too—but those hydrocarbon molecules were shorter. Very short hydrocarbons, which contain up to four carbon atoms, became burnable gases, like methane or propane. Slightly longer ones, which contain five or more carbon atoms, became liquids. They formed different types of oil, which can be distilled into various fuels, such as gasoline or diesel.

So if poop can be cooked into coal, can it also be turned into petroleum?

In fact, it can. As the Ben-Gurion researchers were running their tests in the hot and dry Negev, another team halfway across the world, in the cool and breezy Pacific Northwest, was testing a different sludge-cooking recipe. That team, however, was doing it on a larger scale, and with a lot more money at stake.

From Poop to Petroleum

In 2015, a peculiar shipment traveled across the US-Canadian border in the Pacific Northwest. It was a small collection of five-gallon buckets filled with sloshing sewage sludge. Shipped from Vancouver, the unusual bundle was destined for Richland, a little town of about 50,000 people in Washington State.

The cargo might have been the first international shipment of its kind in either country's history. And this sludge was a rare commodity. It's just not what people—not even biologists and medics—usually send from one country to another. Sludge is something people generally dispose of, rather than ship. And, regardless of the disposal method, it typically stays not only within its country of origin, but usually within the same municipality. The rare exception is the treated, no longer pathogenic, sludge that occasionally crosses state borders when it's destined to become a landfill cover somewhere.

The Canadian sludge buckets were destined for another purpose. Produced by Metro Vancouver's wastewater treatment plant, they were en route to the Pacific Northwest National Laboratory, or PNNL, one of the US Department of Energy's national laboratories, charged with finding and developing alternative energy sources.

"Orchestrating their travel took some time and effort," says Paul Kadota, Metro Vancouver's program manager.

"It required a bit of planning," Kadota recalls. "The PNNL supplied a letter expressing their desire to receive this material, which was part of the package. And from our end, we phoned ahead to say that we are delivering it." That documented handshake ensured the sludge's successful journey. As the buckets crossed the border, the PNNL researchers were eagerly awaiting their arrival. They had big plans. If all went well, this delivery would open a new chapter in biofuels research.

When the sludge arrived, the research team loaded it into their testing apparatus—a complex assemblage of metal tubes, valves, gauges, and chambers. The machine's centerpiece was a sleek, silver serpentine pipe. Inside that pipe, the sludge would be heated to about 350°C and squashed with 3,000 pounds of pressure per square inch, or about 200 atmospheres. In other words, the Vancouver samples would smolder at 100°C hotter than in Ben-Gurion's reactors and under 30 times more pressure than within Cambi cookers. These hellish conditions, akin to those that forged oil and gas deep beneath the seafloor over millions of years, would result in a process called, in scientific terms, hydrothermal liquefaction, or HTL. Unlike the Cambi settings, which mimic regular cooking to make sludge easier for microbes to digest, the HTL environment would break down not only all living cells, but all remaining organic molecules, into shorter, smaller carbon compounds. The hope was that the liquefied mix would burn well, perhaps leading to a new way of making fuel. And Metro Vancouver's sludge was selected as the "guinea pig" for testing the process, Kadota says.

The PNNL scientists fired up their unconventional oven and waited to see what would come out the other end.

The Brief History of Biomass Liquefaction

HTL isn't a new process. Scientists and engineers have been experimenting with it since the 1930s. In fact, a patent dating back to over 70 years ago describes the use of high temperature and pressure to convert wet materials into crude oil or biocrude, but at the time, the technology needed to streamline the process wasn't available. "People were doing it in batch configurations, putting it in tanks instead of pipes and tubes," says Corinne Drennan, who is responsible for PNNL's research portfolio on biofuels, products, and energy. "And instead of biosolids they were trying to use other biomass—like wood."

After the oil crisis of the 1970s, HTL suddenly sparked more interest, and PNNL began experimenting with the process. Over the years, different teams continued working on it, with moderate progress, depending on how generous or frugal their research budgets were. And then, after the financial meltdown of 2008, the concept received a boost.

When President Obama signed the American Recovery and Reinvestment Act in 2009, the Department of Energy received several billion dollars to accelerate the development of various green technologies. One of its projects was to identify the most efficient ways of converting biomass to fuel. The project had ambitious targets. One of its goals was to find the best biomass for the buck; potential candidates included corn, algae, and wood. (Sludge was neither on the list nor on people' minds, even remotely.) Another goal was to identify the most efficient way of cooking plant remains into fuel. Moreover, the process would have to work commercially—that is, it would need to function in a continuous manner, with the inputs easily pumped in and the outputs easily unloaded. The resulting biocrude would then be sent to refineries, because crude oil doesn't make usable fuel—it has to be distilled into various fuel types suitable for specific engines.

The Recovery Act channeled over a billion dollars into two biofuel research groups: one focused on creating fuel from corn and wood, while the other, in which PNNL played a part, placed its bet on algae. Independently from PNNL, a private company, Genifuel Corporation, in Salt Lake City, Utah, had been experimenting with growing algae for fuels for some time. While researching available conversion methods, Genifuel's president, James Oyler, learned about PNNL's HTL experiments and offered his algae for a test. At that time, the PNNL team hadn't yet tried cooking algae into fuel, so they were interested. Oyler scooped up a bunch of green, soupy slurry from his algae-growing ponds and headed to Richland to find out whether his watery cargo could be turned into energy. "I drove to PNNL in a car full of five-gallon buckets of algae," he recalls with a chuckle. "We got it all set up—and lo and behold—it worked great. We made crude oil."

Tapping into the Recovery Act funds, the joint PNNL-Genifuel team secured a grant to study various ways of converting algae into fuel. The project's scope was substantial. It involved not only growing algae and converting it into biocrude, but also identifying the most productive algal species and genetic research to make it into a better biofuel stock. And, of course, the process would have to make economic sense.

After experimenting with various ways to cook the algae, the team found HTL to be the most effective method. "It alone could account for the 85 percent reduction in cost of making algae into fuel," Oyler says. "That got a lot of attention and the interest really took off. The Department of Energy started asking, what else can we do with it?"

That question spearheaded a quest for other suitable feedstocks for the HTL process. Algae, even with the decreased cost, still wasn't a perfect feedstock. It took too much effort and money to cultivate. It needed ponds to sprout in. It needed light to grow, and when it didn't get enough, it would wither. And harvesting it required machinery designed to support industrial-sized operations. Overall, making crude oil from algae was still more expensive than pumping it out of the ground, Oyler explains. Finding a feedstock free of these problems would make the biofuel concept more competitive and commercially viable.

So the question became, what other candidates were there that possessed algae-like properties? What other materials existed that were wet, slushy, and conveniently concentrated in designated spaces—accumulated, perhaps, as by-products of other industries? Ideally, these materials would also be inexpensive—or undesirable, so that their owners would want to get rid of them.

Sewage sludge was a perfect fit. It had all of the above attributes, and it was so unwanted that it had an "avoided cost," as the wastewater treatment industry calls it. There are 16,000 wastewater utilities in the United States, which cumulatively treat approximately 34 billion gallons of sewage every day. That amount of intensely undesirable sludge could produce the equivalent of approximately 30 million barrels of oil per year. That much fuel would need 15 supertankers to transport and would fill up about 40 million Hummer SUV tanks.[1] And while it would make only a small dent in overall US oil consumption, which totals about 20 million barrels a day,[2] it would still help offset fossil fuel use—and would also rid humans of their unwanted fecal matter. What's more, sewage is already conveniently aggregated and separated from people and the environment, not to mention that it is carefully measured, sampled, and monitored.

Oyler reached out to several wastewater utilities to probe their interest. The utilities liked the idea, but none had the people and the resources to vet the process. It was just too expensive for a single utility to test. Testing of the concept required multiple organizations to join forces: the Department of Energy, PNNL, Genifuel, the EPA, and the Water Research Foundation, a nonprofit organization that works to advance the science and technology of water management and helps fund promising innovations. "We sort of scan the universe for ideas," Jeff Moeller, the foundation's director of water technologies, tells me as he explains the foundation's approach. He had originally become interested in algae for a different reason—he attended a workshop on using it for cleaning wastewater, which eventually led him to Genifuel. "So we kind of stumbled upon this." The Water Research Foundation brought together 10 utilities, each of which pitched in $5,000, adding to some money from an EPA grant. One of those 10 utilities was Metro Vancouver, which later sent samples from its wastewater treatment plant across the border.

The 15-Minute Cinderella Makeover

As the Vancouver sludge inched along inside the serpentine pipe, the high temperature and pressure began to work their magic. The molecular bonds broke. The atoms separated. The chemical elements realigned and reconnected, forming new compounds. The biological

materials followed the same metamorphoses as the organic deposits on the bottom of the ocean, only faster. Within a 15–20-minute
span, the sludge underwent a truly Cinderella-like makeover. It transformed from an off-putting muck into a highly valuable resource consisting of three distinct components: biocrude, water, and gas. Plus,
the biocrude appeared to be neatly separated from the water—the
output looked like oil and vinegar poured together for a salad dress-
ing but not yet mixed. The biocrude was essentially oil—the black
liquid gold the world runs on. And it looked like it could be sent to
a refinery right away.

"It was a pleasant surprise for us," Drennan recalls. "Sewage was
better than wood and other types of biomass." Some biomasses contain too much oxygen, which is corrosive and interferes with the
production process. Others release too much acid, which yields a
bunch of gunk. "If your system is too acidic, you end up condensing
everything and just making a goo instead of the biocrude," Drennan
explains. Sewage sludge had none of these problems.

Like the Ben-Gurion team, the PNNL scientists discovered the
unexpected benefits of fat floating in the sewage. From consumed
but undigested fats to grease poured down the drain, these oily
substances were a boon to biocrude production in more ways than
one. "Fats are good lubricants. They facilitate the sludge movement
through the HTL conversion tube," Drennan says. "And they also
burn well—because of their chemical structure. If you think about
fat molecules, they are triglycerides or free fatty acids—long chains
of hydrocarbons that have a little bit of oxygen at the end. They are
probably the closest biological molecules to the inorganic hydrocarbons. They are almost diesel molecules already. And they seem to
keep the sludge moving through the reactor and facilitate the conversion of other wastewater materials, like toilet paper. That makes
them conducive to producing a very high quality biocrude that, when
refined, yields fuels such as gasoline, diesel, and jet fuels."

These results were also a pleasant surprise for Metro Vancouver.
"When we came over to PNNL after the testing was done, there were
at least a dozen researchers in the room, all looking very happy and
waiting to shake our hands," Kadota recalls. "They were all scientists,

so they were all about the numbers—and they were excited about the numbers our samples produced." The Water Research Foundation was equally thrilled. "There was a lot of energy and excitement at that meeting," Moeller recalls. "The quality of the biocrude oil produced from wastewater sludge was as good or even better than biocrude produced from the best strains of algae!"

The Metro Vancouver team decided to build a small test HTL installation at one of their treatment plants to assess how well the process could work on a commercial level. But building even a small system that would pump sludge into a steel serpentine pipe and spit biocrude out the other end was an expensive proposition. Kadota's team wrote a proposal to the Metro Vancouver Board asking for 4 million Canadian dollars to build the test installation. The board was willing to shell out C$4 million, but it wanted to have another big player in the game who would be willing to shell out even more—C$5 million. The project just needed a finishing touch—and it came from an unexpected source.

Black Gold, Refined

Kel Coulson doesn't remember exactly when she read about the process of converting poop to petroleum. She doesn't remember exactly what the write-up said, either. But she remembers reading it and thinking, "Can this be real?"

An environmental engineer by training, Coulson worked at the Burnaby Refinery in British Columbia, owned at the time by Chevron and currently by Parkland Fuel Corporation. Like many other carbon-heavy businesses, the company was looking to reduce its carbon footprint to meet government-set goals. One of the ways it could do so was by supplementing its nonrenewable crude oil with some renewable content. The American government has a similar program: any company that sells fuel has to make a certain percentage of it from renewable stocks—otherwise it must pay a fine or buy so-called credits from "greener" businesses. Coulson's company was experimenting with adding canola oil to its feedstock, which in industry terms is called "co-processing." Canola, however, wasn't ideal.

It's a food item, Coulson points out, and should be used as such, rather than distilled into diesel. Waste products would make much better candidates—and so Coulson was looking for suitable wastes.

The post on her LinkedIn feed popped up sometime in 2017. She recognized the name of the author—Paul Kadota from Metro Vancouver. Coulson knew Kadota well because she had also worked in the wastewater sector. She picked up the phone and called him. "Does it really work?" she asked. "Because if it does, we are interested."

Greening up fuel was becoming a hot topic. "Our customers, including long-haul transportation in marine or aviation industry, are looking for greener fuel to meet their climate targets," she explains. For example, the International Maritime Organization is tightening pollution control, so ships that traditionally burned heavy bunker fuel will have to start switching to diesel. "As a fuel producer, we need to produce more diesel," Coulson explains. "And not just any diesel, but diesel with a higher renewable content. So we have a high interest in this idea."

Kadota had the waste technology ready, and Coulson's company was willing to help with financing. "To be able to repurpose this carbon source into a high-value product is really a win for his industry and our industry," Coulson says.

With a total budget of C$9 million, Kadota's team sketched out a plan and a timeline, adding the HTL piece to their plant's infrastructure in a way that wouldn't cause any service disruption to Vancouver residents. Altogether, Metro Vancouver's five sewage treatment plants process the contributions of 250,000 people. The HTL installation would be "plugged into" the Annacis Wastewater Treatment Plant, diverting a small sewage stream from about 30,000 people into the serpentine pipes. The team planned to finish the design phase by mid-2020, manufacture the unit later that year, and produce the first biocrude in 2021 or 2022. That biocrude will be trucked from the plant to the refinery, the same way oil is delivered to it today. If that works well, the process will be scaled up, and Vancouver residents will be powering their cars—and possibly boats and airplanes—with their own organic power. Legacy combustion engines won't fully disappear anytime soon, so running them on renewable biofuel is a good alternative to fossil fuels. "It's one of the most favorite projects of my career!" Coulson says.

In addition to the obvious paybacks—solving the sludge problem and making renewable energy—converting poop to petroleum has other benefits. For starters, it completely eliminates the yuck factor and leaves no hygienic concerns for even the most fastidious consumers.

Conversion of sludge to petroleum also solves the issue of long-haul transportation. When converted to fertilizer, sludge must end up on farmland, and therefore has to be transported to the countryside—farmers don't want to travel to get it, even when they can take it for free. But poop-derived gasoline doesn't have to travel anywhere. It can be used right next to its original source and *by* its original source—in this case, Vancouver residents. Moreover, it has the potential to make wastewater utilities energy-neutral. "We've done some research that shows that there is five times more energy embedded in wastewater than what's needed to treat it," says Moeller. "So energy is a really exciting component of it."

HTL has yet another enormous benefit, which solves a pressing problem of modern sewage: the abundance of synthetic chemical compounds within it. Unlike the product of the sewage farm days, contemporary sludge is brimming with hormones, antibiotics, and other pharmaceuticals. It also includes other chemicals poured or tossed down the drain, such as pesticides, paints, plastics, and even fire retardants that seep in from laundered fabrics. The HTL process breaks down all of them because no molecular bonds can survive its heat and pressure.

Finally, HTL allows for the filtering of useful ingredients from the sewage—such as phosphorus. The PNNL team used a filter to gather phosphorus from the forming biocrude, so that this important fertilizer could be returned to the land. Other useful sewage ingredients, including precious metals like silver, gold, and palladium, can be recovered, too—and technologies that can do it are already in the works. Sewage treatment companies are literally sitting on a gold mine—or perhaps a rare metal mine—and they are beginning to appreciate it.

"The utility companies are realizing that wastewater can be a valuable resource and there are a lot of things we can recover from it—including energy, nutrients, and chemical compounds," says Moeller. Some researchers are even looking at other options, such as producing bioplastics from sludge, he adds. "What's really exciting is that there is a paradigm shift!"

But there is also another paradigm shift brewing, in the very field that discovered and taught us about the dangers of our fecal matter: medicine. For over a century, medics have warned humankind to stay far away from its own waste. Now they are finding that it has powerful medicinal properties that can fight off superbugs, boost the immune system, and possibly even cure autoimmune diseases.

Yes, poop can kill you. It can also save your life. And that is perhaps the most counterintuitive revelation of all.

PART 3
THE FUTURE OF MEDICINE AND OTHER THINGS

CHAPTER 14

POOP: THE BEST (AND CHEAPEST) MEDICINE

Catherine Duff was dying. The raging infection in her intestines had been slowly eating her up from the inside out for weeks, leaving her unable to digest food, absorb nutrients, or move around her house. She hadn't left her bedroom for days, except for bathroom trips due to continuous diarrhea. Worse, it wasn't the first time this infection had brought Catherine close to death. She wasn't looking forward to repeating that experience.

Catherine, a 50-something mom and grandma in Indiana, was battling *Clostridium difficile*, a much-feared intestinal superbug that afflicts nearly a half million Americans a year, killing close to 30,000 of them within 30 days, according to statistics from the Centers for Disease Control and Prevention. It is considered a health care–associated infection—meaning that it disproportionately affects people staying in hospitals, undergoing surgeries, and taking antibiotics. This bacterium, nicknamed *C. diff*, causes pseudomembranous colitis, which manifests itself as continuous diarrhea, dehydrating patients to the point of kidney shutdown, which Duff had already gone through once. That time, a neighbor had found her unconscious and rushed her to the emergency room. She woke up in the intensive care unit.

C. diff is a very stubborn blight. It responds only to antibiotics

The original version of this story, "The Magic Poop Potion," appeared in *Narrative.ly*, July 30, 2014.

of last resort—such as vancomycin—and it takes months to clear. During Catherine's first five bouts with *C. diff*, vancomycin worked. It cost her insurer over $200,000 in one year, but it kept her alive. But in April 2012, during its sixth reoccurrence, her *C. diff* became antibiotic-resistant, and even vancomycin stopped working, making Catherine realize that she might not live this time. When she saw that modern medicine had nothing left to offer her, she and her husband, John Duff, decided to try something so radical it would make most medics in the world shudder.

Catherine and John decided to try a fecal transplant, which in layman's terms amounts to taking a healthy person's poop and putting it up a sick person's rear. The couple weren't going to try it as part of an experimental study. Nor were they planning to go to a qualified research center or a medical office. They were going to do it in their own home, in their own bed, and with their own, pharmacy-bought plastic enema kit. They had read on the internet that some *C. diff* sufferers had cured themselves with fecal transplants, and they were going to follow the same exact steps those people did. Catherine's odds of survival were diminishing so rapidly that she had little to lose. Now she lay in bed waiting for John to return home with all the necessary equipment, which included an enema kit, cheesecloth, a funnel, and a device called a hat. When she heard him come in, she was ready for the experiment.

"Honey, I'm home," John called from downstairs. Then he went to the kitchen to cook up a recipe that might save his wife's life.

All Prejudices Come from the Intestines

Following a miracle cure recipe from the internet may not sound like a smart approach to treating an antibiotic-resistant superbug, but it is a logical one. According to the Human Microbiome Project—the National Institutes of Health's initiative to understand how microbial communities affect health and disease—microbes populate the outside and inside of our bodies in such abundance that their cells outnumber ours by about ten to one. Many species dwell on our skin and in our mouths, but the vast majority make our intestines their home, where they perform many important functions. They help us absorb nutrients in our food by breaking it down into simpler, more easily absorbable compounds—about 10 percent of our calories come from microbial digestion. They synthesize certain nutrients without which we wouldn't be able to function. They send chemical messages to our brains, affecting our moods and neural responses. In a similar way, they interact with our immune systems, mediating our responses to pathogens. They are also quite territorial. "All prejudices come from the intestines," wrote philosopher Friedrich Nietzsche—and he wasn't that far from the truth. Our microbial communities form colonies in our intestines, and they aggressively attack and kill foreign invaders, destroying many pathogenic species and preventing infections before they take hold.

The bacterial menagerie populating our guts, collectively known as the intestinal microbiome, is directly related to our well-being, so having the right combination of microbes in one's belly is indeed key to one's health. Microbiome deficiencies or abnormalities have been linked to many modern maladies plaguing humankind, including asthma, celiac disease, immune deficiencies, and autoimmune disorders. Scientists have begun to realize that our intestines are like a rainforest inhabited by trillions of creatures that form complex ecosystems, which keep pathogens at bay.

Studies found that people with obesity and diabetes shared similar microbiome deficiencies. People with persistent intestinal conditions, such as Crohn's disease or irritable bowel syndrome, had decreased populations of certain bacterial species. It turned out that people with Alzheimer's disease had less diverse microbiomes than healthy people of the same age. One group of researchers found that Parkinson's disease might also have its origin in the gut, which may open up novel possibilities for treatment. Another team compared the amounts of various microbial by-products in the feces of healthy children and of children with autism—and found that 50 of these substances differed significantly between the two cohorts.[1] A later study suggested that microbiome therapies might be a safe and effective treatment for autism spectrum disorder.² Multiple studies linked autoimmune disorders, the numbers of which have been skyrocketing, to microbiome deviations as well.

When we take antibiotics, the medicines don't kill only the pathogenic bacteria, but inevitably some of the beneficial ones as well. Studies found that patients' microbiomes dwindle after antibiotic treatments. Usually they rebound after a while, as the gut is repopulated with similar species. But prolonged antibiotic use can decimate microbial gut communities so profoundly that they can no longer defend against pathogens.

Furthermore, different antibiotics work differently. Some are milder, so they affect only a limited number of beneficial strains. But stronger ones target a wide variety of species, so while they help eliminate a greater range of pathogens, they also wipe out many different beneficial organisms. So when patients take very strong antibiotics, like vancomycin, to treat stubborn or recurring infections, and do so repeatedly or for a long time, their overall intestinal microbiome declines greatly. Alexander Khoruts, a gastroenterologist at the University of Minnesota, and one of the fecal transplant's early pioneers, calls such antibiotic treatments "carpet bombing," after which "all the normal stuff is gone." And its absence paves the way for *C. diff* to return. Extinguishing *C. diff* takes even more antibiotics, which further damage the intestinal microbiome.

That's why health care–related superbugs like *C. diff* selectively afflict hospitalized patients. These people become easy targets for that opportunistic pathogen. Their microbiomes and immune systems are already so weakened by their existing health issues, surgeries, and antibiotics that it's easy for the pathogen to take hold. That's why, while *C. diff* is deadly, it's not highly contagious. When patients with *C. diff* are sent home from the hospital to be cared for by their family members, the latter typically don't fall ill—because they usually have normal, healthy bacterial communities that effectively guard against this infection. Despite the fact that John Duff shared the same kitchen utensils, the same bed, and the same bathroom with his wife, he never got *C. diff* because his intestinal microbiome hadn't been damaged by antibiotics.

Killing a microbial community is easy. Restoring a damaged microbial community is much harder. Medics don't yet know how much of what species should be present in one's gut to ward off *C. diff* or other blights, so they can't yet give patients a prescription pill with the right bacterial combo in it—although they are working on it.

Under these circumstances, the idea of a fecal matter transplant (abbreviated FMT) begins to make sense. The concept of replacing a sick person's microbiome with a robust microbiome from a healthy person—by transferring the bacterial community in its native habitat—could be life saving. It really isn't much different than a blood transfusion—only instead of blood cells, you are infusing the person with an active and healthy microbial zoo. So FMT could indeed be Catherine's cure—if done correctly, with sterilized equipment, and by a medical professional.

The last requirement, however, proved to be a problem. The microbiome transplant paradigm hadn't yet taken hold in 2012. At the time, the prevailing doctrine still viewed human fecal matter as the epitome of disease. And while the human microbiome was already being intensely researched, FMT was far from a mainstream procedure. Catherine's daughter read about it on the internet, but the only two doctors performing it were in Nevada and Australia, and Catherine was too sick to travel that far.

The Duffs couldn't find anyone closer. Doctors didn't think FMT was curative, or for that matter, ethical. They were afraid of complications. Most of them had never even heard of it before, let alone administered one. Catherine's own gastroenterologist wouldn't do it, but luckily, he agreed to screen John to be a stool donor—to ensure that he didn't have any diseases or pathogens he could pass on to her. A week before the couple attempted the procedure, John sent his stool sample for a test and was declared pathogen-free.

The Blender Returns

On the day of the transplant, the couple wasn't nervous. "We were ready for it," John recalls. "We wanted to move on." John's take on the procedure was simple and practical. "Yes, it involved some things I've never done before, but they weren't extraordinary. I was comfortable with what I would have to do."

In his bathroom, John took the first piece of transplant equipment out of its vacuum-sealed bag and placed it over the toilet. The device, sterilized and germ-free, was needed to avoid any possible contamination while collecting a specimen. Called a "hat" in medical lingo, the device looked like one, too—its white brims kept it on the toilet while John was delivering the specimen into its basin. "That part of it was somewhat routine," he remarks. When he was done, he took the sample-containing hat to the kitchen.

Lystek's creators weren't the only people who thought of homogenizing feces for easier use. Fecal transplant patients had arrived at the same idea. They had a similar problem with their medium: it was too thick and not homogenized enough to be pushed through tubes, never mind the very thin openings of an enema bottle. The instructions the couple found online suggested mixing the donor's stool with the contents of an enema bottle, then pureeing the mixture into a mush in a kitchen blender.

Still careful to avoid bacterial contamination, John sterilized all the equipment first. He poured boiling water over the blender, funnel, and cheesecloth. Then he sloshed his sample into the blender, emptied the enema bottle into it, and pushed the button.

The blender roared and whipped the mixture inside the glass, fusing it into a dark-brown emulsion. John's blender was a regular inexpensive home model—unlike Lystek's creators, he didn't need to shear bacterial cells, but rather wanted to preserve them in their entirety. Yet he needed the mixture to be uniformly fine and fluid to pass through the enema kit's nozzle, so he had an extra step to do.

When the fecal matter was blended well enough, he stretched the cheesecloth over the funnel, poured his cocktail onto the cheesecloth, and sieved it through. When it was necessary, he smashed the remaining glops into mush to make sure they wouldn't clog the nozzle. "I mashed it through because it's not something that flows readily," he explains—calmly and matter-of-factly. He doesn't remember whether he used gloves in the process or just did it with his bare hands. But if he did wear gloves, it was not to protect himself from his own poop, but to avoid accidentally transplanting unwanted germs. "I may have used gloves, I probably did, I was very careful to maintain a sterile environment," he explains. "It wasn't to prevent the sample from getting on my hands, but to prevent my hands from getting on the sample."

John was completely unfazed by the process because he had had good training. He was a former submarine officer who had spent months at sea. Being inside an underwater vessel makes one immune to the smells of human excrement, he explains. When a submarine waste tank is emptied, the contents are expelled under pressure—and then the very air used to create that pressure is sucked back in— because there's only so much air in the submarine. That recycled air brings smells with it, smells that waft through the vessel's compartments multiple times a day. You get so used to it, you just aren't bothered by it anymore. "It looked like a chocolate milkshake," he says, describing his transplant mix in exactly the same words Lystek's creators used. It didn't smell like a milkshake, but John didn't care. He screwed on the top of the enema bottle and carried it upstairs to his wife.

By now Catherine was a veteran patient. She had spent the past 20 years battling a slew of health problems stemming from an unfortunate accident in her youth. "I used to be a very active person," she says. "I ran, I hiked, I swam, I skied, I canoed, and I rode horses all the time." But when she was 35, she fell off a horse—and the horse fell on top of her, breaking her leg into 12 pieces. Catherine went through 17 surgeries to put her leg back together, but in the process she lost about three centimeters of bone length. When her leg finally healed, it was much shorter than the other one, and Catherine ended up with a limp. Besides the aesthetics, the limp brought other problems: it would eventually wear out the cartilage in her knees. "I knew I would have to do a knee replacement surgery later in life," Catherine says— but more problems were to come. Years of uneven walking would eventually damage her hips and shift her spinal vertebrae, pinching her spinal cord in between them. All that meant many more surgeries to come.

That wasn't all. As she grew older, Catherine developed diverticulitis, a condition in which little pouches form inside the colon. Often, these pouches cause no issues, but when they become inflamed—for example, when nuts or seeds get stuck inside them—they require antibiotic treatments and, in some cases, surgery. Catherine happened to be that extreme case: in 2005, her intestines became so inflamed, despite weeks of antibiotic treatment, that part of her colon had to be taken out. The surgery had complications: Catherine developed an abscess that burst, causing a systemic infection—and her doctors put her on very strong antibiotics. While still in the hospital, she contracted *C. diff* for the first time, and it took months to clear. When she went back for a follow-up surgery, *C. diff* returned—for several more months. "I ended being very ill for almost a year, almost died a few times," she recalls.

But her ordeal was only beginning. From that point on, *C. diff* came back every chance it got—and unfortunately, Catherine's condition offered it many chances. In 2007, she had a hip procedure to address the worn-out cartilage—and within two days, the plague returned. Two years later, another hip procedure brought it back for the fourth time. The next year, a knee replacement invoked it again. After that, even the mildest dose of antibiotics, which are sometimes required even for dental work, would bring another round of *C. diff*. "My orthopedic surgeon and dentist assured *C. diff* won't return, but it took one dose of amoxicillin and it did within a few hours," Catherine says of yet another occurrence. "Every time it happened, it took months to go away. And after every bout of it, I got weaker and weaker. It takes months to recover once *C. diff* is gone, just to get your energy levels back. And I never got my energy levels back to where I was before."

In July 2012, she had to have emergency spinal surgery because of a pinched nerve—and sure enough, she woke up with *C. diff*. She was worn out from constant infections and poor food absorption, and no drug combination was working. "I was so depleted my immune system was at zero, it was destroyed."

Her colorectal surgeon wanted to remove her colon entirely and replace it with a plastic bag—that is, to perform a colostomy, a procedure that essentially leaves one with an intestine outside the body.

But Catherine was so weak, he wasn't sure she would survive the procedure. Catherine didn't want to do it either. Even if she survived, she says, she wouldn't call it "living"—she had had the bag temporarily during her abdominal surgery and it was a horrible experience for her. "Every time you go to the bathroom, it comes out of the bag," she says. "You can't go anywhere because you don't know when the bag is going to make noises or leak."

Compared with that, a homemade fecal transplant felt like a breeze.

When John came upstairs with the milkshake, Catherine had no qualms or hesitations. She turned onto her left side, with her knees toward her chest, as the instructions prescribed. John smeared some lubricant onto the nozzle and gently inserted it. He pressed on the bottle's rubbery body, but felt no movement. A thought—"Did I make it too thick?"—flashed through his mind. It wouldn't be hard to fix, but they both wanted to get it done already. John adjusted the angle and squeezed harder—and the milkshake began to flow. He got about 80 percent of the bottle in—which should have been enough.

Catherine turned onto her back and lifted her legs up, slowly turning her body from side to side to spread the mix inside her. The instructions suggested elevating the foot of the bed to make the transplant flow up the intestines as far as possible. So John had already propped the bed's end on a six-by-six piece of lumber from his garage. "The enema reaches only your lower colon, but the infection is throughout your colon so you have to try to make the sample spread," he explains. "And then you are supposed to keep it inside you and not go to the bathroom for as long as you can."

Once all was done, John gathered his homespun transplant kit—funnel, hat, cheesecloth, blender, and enema—and sealed it all in a hazardous waste disposal bag. And then they both waited for the matter to work its magic.

And magic it was. Four hours later, when Catherine got up to go to the bathroom, she felt different. Something inside her had changed. The diarrhea stopped. That evening, she slept through the night for the first time in months, without having to run to the bathroom multiple times. The morning brought even more pleasant surprises. "When I woke up, I felt good enough to take a shower and dry my hair and put on makeup and jewelry—things I hadn't done in months," Catherine recalls. "Within two days she was driving a car," John says—after months of not leaving her bedroom, let alone the house. *C. diff* was completely gone, and Catherine felt like a new person.

In fact, she felt so good that she wanted to do something for all other *C. diff* sufferers who weren't as lucky as she was in their efforts to heal themselves. She decided to launch a Fecal Transplant Foundation to make it easier for people to find out about this simple miracle cure.

The Hidden History of the Fecal Transplant

It may sound as if the Duffs stumbled upon a novel, experimental, and previously unknown treatment forged by some obscure but brilliant bacteriologist. But that's actually not true. Transplanting healthy people's feces into sick ones to cure lethal infections is one of the oldest remedies documented in writing.

During the Dong-jin dynasty in the fourth century, a well-known doctor of traditional Chinese medicine, Ge Hong, described a recipe for a liquid fecal therapeutic for patients who suffered food poisoning or severe diarrhea—whether from *C. diff* or something else, there's no way of knowing. The sufferers had to swallow the mix, which, despite its yuck factor, did wonders. It literally brought people back from the brink of death and was considered a medical miracle. Reported in the first Chinese handbook of emergency medicine, *Zhou Hou Bei Ji Fang*, or *Handy Therapy for Emergencies*, the method is the earliest documented form of fecal transplantation.

The method worked so well that more sophisticated concoctions were developed later. In the sixteenth century, Li Shizen, a famous Chinese medic and a revered polymath, much like Leonardo Da Vinci in the West, described not one but a series of fecal preparations in his book *Ben Cao Gang Mu*, or *Compendium of Materia Medica*. The prescriptions were quite elaborate. They included fresh fecal mixes, dry feces, infants' feces, and even fermented fecal solutions used to treat abdominal diseases characterized by severe diarrhea, fever, pain, vomiting, and even constipation. Chinese doctors made these therapeutics without sterilized hat devices, electric blenders, or plastic enema kits. And they used their own recipes, says Liping Zhao, director of the Laboratory of Molecular Microbial Ecology and Eco-genomics at Shanghai Jiao Tong University.

One recipe, for example, involved dropping a hollow bamboo stick into an outhouse pit and leaving it for the winter. The material accumulated inside would then be taken out and dried into a powder. The sick would take the powder by mouth. Another technique involved putting babies' feces in a jar and burying it in the soil for three years, where it would presumably ferment. The most elaborate formula prescribed mixing a healthy adult's waste with clay and burying the jar in soil for 10 years, during which it would turn into an odorless, viscous, clear yellow liquid described as "golden juice."

What all three methods shared was that the fecal matter was kept under anaerobic conditions for some time, whether in outhouses or buried jars—which may have kept out aerobic pathogens and fostered the growth of some anaerobic bacteria that should be present in human guts. That may have indeed given the concoctions medicinal qualities. "They were fascinating production methods," says Khoruts. "It would be very interesting to know what they came up with," he adds, but so far no one in the Western world has attempted to re-create these potions.

None of these methods is officially practiced in China today, Zhao says, but some traditional doctors may still use them. In fact, a few years ago, a construction company accidently dug up a reservoir of buried golden juice jars while excavating near a Buddhist monastery. "Buddhist monks have a very healthy intestinal flora, so they collect and bury their fecal matter every year," Zhao says. "And every year they dig some out and give it to people for free."

There are other occasional mentions of fecal transplants throughout history. Supposedly, in the seventeenth century, Italian anatomist Fabricius ab Aquapendente used it to treat gastrointestinal diseases in veterinary medicine. But the first documented modern doctor who tried fecal transplants on humans was American surgeon Ben

Eiseman. In 1958, he gave fecal enemas to four patients who suffered pseudomembranous colitis after a high dose of antibiotics. Today, pseudomembranous colitis is recognized as a severe case of *C. diff*. Back then, it hadn't yet been described as such, but researchers already understood that it was related to antibiotic use, potentially due to a lack of normal bacteria in the colon. Eiseman reasoned that antibiotics had killed the normal intestinal bacteria, so replenishing them would solve the problem. He turned out to be right. Normally, only a quarter of patients with pseudomembranous colitis survived it, but those treated with fecal enemas recovered within days.[3] However, shortly afterward, vancomycin was discovered to be an effective treatment, and fecal transplants were once again largely forgotten.

Then, in 1989, Justin Bennet, a medical student who later became a gastroenterologist, gave himself a fecal transplant to cure persistent ulcerative colitis, a type of chronic inflammatory bowel disease, or IBD. Bennet had suffered from ulcerative colitis for years. The refractory condition responded to medications poorly, and as soon as the drugs were stopped, it would flare up again. When Bennet gave himself several high-volume enemas, filling his gut with flora from a healthy donor, his colitis all but cleared. For the first time in 11 years, he didn't need medication, and his intestinal tests showed no sign of active infection. When he crossed the six-month no-meds mark, he wrote a letter to *The Lancet*, which was published as a de facto account of this unconventional therapy.

In the meantime, around the 1990s, physicians started noticing more aggressive forms of *C. diff*, which some doctors also called CDI. More virulent strains emerged that propagated more efficiently, produced greater quantities of toxins, and resisted multiple antibiotics. And even when vancomycin killed the bacteria, their spores managed to survive in some patients—so the infection would return as soon as the antibiotics were stopped. As a result, by the early twenty-first century, *C. diff* had become so deadly that it was killing more people than HIV infection. "The US Centers for Disease Control and Prevention estimates this infection kills about 30,000 people per year in the United States alone, and they admit it is a conservative estimate— more realistic estimates are over 100,000 people a year," Khoruts reported in his 2014 interview with *Global Advances in Health and Medicine.* "To give some sense of proportion, CDI in the United States is far bigger in terms of mortality than AIDS."[4]

Khoruts performed his first fecal transplant on a patient with *C. diff* in 2008. Several other gastroenterologists, including Australian doctor Thomas Borody, also tried it as a measure of last resort—nearly always with similarly miraculous results. Serendipitously, in the same year that Catherine did her own transplant, four scientists wrote a letter to the *American Journal of Gastroenterology,* titled "Should We Standardize the 1,700-Year-Old Fecal Microbiota Transplantation?"[5] The letter described the fecal transplant's long history and noted its consistent efficacy. The authors, some of whom worked at the hospital of Nanjing Medical University and some at the University of Maryland, wrote, after reviewing the recent medical literature, that transplant efficacy reached 98 percent.

Standardizing the treatment, however, would mean involving the US Food and Drug Administration, which would draw up rules and regulations for fecal transplants after putting the process through rigorous testing and noting all potential side effects.

Exactly how easy would it be to get the FDA to approve poop enemas?

Catherine Duff was about to find out.

The Battle to Save the Fecal Transplant

In May 2013, Catherine drove up to the security gates of the National Institutes of Health in Bethesda, Maryland. She was there for an important event. A year after she had won her hardest *C. diff* battle, the FDA was hosting a public workshop titled "Fecal Microbiota for Transplantation." For two days, doctors and policymakers from the FDA and Centers for Disease Control were planning to discuss the risks and benefits of the procedure that had saved her life, and Catherine was excited. She had recently launched her Fecal Transplant Foundation to help patients find doctors who could test stool donors and possibly administer the procedure in medical settings.

At the gates, she was greeted by four armed guards and dogs

on leashes. They asked her to get out of the car and proceed into a booth for a security check that she describes as "the strictest . . . I've ever been to in my life." The guards patted her down, emptied her purse, and examined every item she had in it, while the dogs sniffed through her Chevrolet Tahoe looking for explosives. Catherine, who describes herself as a shy, timid person, felt somewhat intimidated.

When she finally walked into the Lister Hill Center Auditorium, where the event was going to take place, she felt intimidated again. She realized that out of the 150 participants in the FDA's public workshop, she was the only member of the public. Everyone else was either a doctor, a researcher, or an affiliate of the CDC, FDA, or another such institution. Every other attendee sported either a PhD or MD on their badge, while she had neither. Most attendees—men and women—wore business suits, while she was dressed far more casually. She felt out of place and uncomfortable, so she found a seat far away from the auditorium's front rows and settled in the bleachers, with her iPad ready for note-taking.

She believed that the FDA's workshop meant that this unconventional treatment was on its way to becoming a mainstream procedure that would save about 30,000 lives every year. She thought that the method was finally getting the attention it deserved, and looked forward to learning all about FDA's the new policies.

But that's not what happened. After the first few presentations, Catherine realized that the procedure that had saved her life was itself in danger. As word about the therapy's success spread, the FDA was being pressured to regulate it. So the FDA had called the meeting not to spread awareness of the new treatment that could save thousands, but rather to rein in who could perform it, where, and why. That would severely limit patients' already restricted access to FMT.

There were reasons for regulating the treatment. Health problems linked to the microbiome—such as autoimmune and cardiovascular diseases or diabetes—could be passed from a donor to the recipient. Other long-term consequences could develop, too. The FDA representatives were concerned that the outcomes of FMT had not been studied enough to make the procedure a standard of care.

The agency's representatives, led by Jay Slater, director of FDA's

Division of Bacterial, Parasitic and Allergenic Products, explained that they were going to label the fecal transplant an "investigational new drug," or IND. Any doctor who wanted to perform it would have to file an IND application and get the FDA's approval.

"The process of doing that sounds very logical, but is extremely arduous," says Khoruts, who attended the same workshop. He had had a fecal transplant IND approved before and could attest that it required an extraordinary amount of work. "I put that application on the scale just to weigh it—it weighed twenty-two pounds. It had to be packaged in several boxes and took us a year to complete." He clarifies that "us" was a university team of 10 researchers. A single practitioner in a small office would never be able to get it done. And no patient withering from *C. diff* could wait that long. "To treat an occasional patient with this procedure, you're not going to fill out an application that size," Khoruts says. "It basically meant that FDA shuts it down for most people."

But at the same time, some FMT researchers were reporting such incredible successes they felt that treating patients with antibiotics was downright wrong. Colleen Kelly, a gastroenterologist from the Warren Alpert Medical School at Brown University, presented a study that compared treatment outcomes of *C. diff* patients treated with fecal transplants with those of patients treated with antibiotics. The transplant cohort had a 98 percent success rate, which was so superior to that of the second group that the researchers halted the study entirely. They felt it was unethical to continue giving people antibiotics. Those people needed transplants instead. And they certainly didn't need a massive amount of paperwork.

Catherine didn't know that the required IND application weighed more than her baby grandson, but she grew more frustrated and angrier with every presentation she heard. How could the agency be so concerned with possible future complications—which overall sounded manageable and even treatable—when people needed the transplant to save their lives now?

"This is what was getting me so riled up," she recalls, quoting the FDA's main argument. "'We don't know the long-term consequences of this, we don't know the long-term consequences of that.' But here

I am, the person dying from *C. diff*, and I am not responding to antibiotics, so what are my options? I will take my chances of developing diabetes or cardiovascular disease because it may or may not happen in the future—but right now I am dying and I want to live!"

She wanted to stand up at the mike and tell all those people in suits what it's like having *C. diff*—running for the bathroom every two hours, passing out from dehydration, and not knowing whether you'll live until tomorrow—but she didn't know how to do it. She had been afraid of public speaking since childhood. In high school, she dreaded presenting reports to her class. At work, she got panic attacks when she had to speak in front of a small group of co-workers. And why would all these experts listen to her, anyway?

The second day of the workshop rolled around, and the same theme continued. Catherine kept waiting for one of the pro-transplant medics to finally counter the agency's regulatory push—but no one did. The word "IND," the concept of which she wasn't fully understanding, was repeated more and more often. It occurred to her that if she didn't do something, the fecal transplant cure would die in its infancy. But what authority did she have among all these doctors and professors? She wasn't just the only member of the public at the workshop, but the only one on the receiving end of the agency's regulatory change. And she realized that if she didn't speak up, no one would.

"They kept saying that only doctors with the IND were going to be able to do the transplants," she says. "I did know enough to understand that it meant more people were going to die." She decided she was going to speak up—for all those patients who would be affected by the new IND rules. "I was really mad and really scared," she says. "But I felt this huge responsibility for all these other people, and that made more mad than I was scared."

Over lunch, she found a quiet spot in the cafeteria and huddled over her iPad, typing up her feverish thoughts into some coherent statements. Then she walked over to the panel's moderator. A six-foot-tall man in a suit, he towered over her petite frame, but she told him she wanted to speak.

The man countered that only medical professionals were allowed to speak at the mikes—but she could ask a question instead. Catherine's eyes swelled with tears. "This is supposed to be a public workshop," she said, shaking. "But look at the attendee list—I am the only member of the public here. No one is talking about us, patients. I am the only patient here and I have to speak!"

The moderator, shocked by her tears and the intensity of her voice, relented. He told her to sit in the microphone row and raise her hand after the next panel's second speaker was done.

She did. She sat down, hugged her iPad, and tried to battle a full-blown panic attack. She was afraid of messing up her speech, of slurring words and not making any sense. She didn't hear anything that the next two speakers said—until the moderator motioned to her to stand up at the mike.

Every attendee was now looking at her. Her heart pounded in her chest as she got up. She thought she was going to faint. Or worse—throw up from all the nervous tension inside her.

Catherine swallowed a lump in her throat, and began to talk.

"I seem to be the only actual member of the public that's here at the public forum," she began reading from her screen. "I'm one of those people who call and email you every day . . ." She tried to speak in a calm and measured way, but as she went on to describe her story, her desperation, and her miraculous recovery, the emotions took over, and she broke into tears. The letters on her iPad blurred. The thoughts in her head jumbled. She lost her place in her speech—and just went on speaking from her heart.

"If your spouse, child, parent, sibling, or best friend were dying from antibiotic-resistant *C. diff*, I imagine that all of you would want them to be able to try FMT," she told the audience. "People are dying every day, today, right now. I have a wonderful husband, three amazing daughters, and two small grandchildren, and I want to live. All of us just want a chance to live. Please, do something not only for me, but for all those around the country and everywhere who have no insurance, no financial resources, no computer with which to Google information, and no hope. Please do something quickly."

She finished and stood there with tears still streaming down her face. Through a blur, she saw Slater, the FDA's director, gesture to her to look around. She was not the only one crying. Other attendees and panelists sat there wiping away tears, too. Khoruts described it as a very deep, moving moment. "There was a brief moment of silence," he says, "and then the whole audience burst into applause."

And then all hell broke loose. The workshop's structured official atmosphere went puff. "People were standing up and talking, and raising their hands and talking, and people were talking to each other in the audience, and asking specific questions of the people on the panel," Catherine recalls. And they were asking very pressing questions.

"We don't have three years to wait and there are a lot of patients like the one who just spoke," said Bob Orenstein from the Mayo Clinic.

"How are we going to help these patients while studies are going on?" echoed Johan Bakken from the University of Minnesota.

"We don't have a clinical team to go out and do all these trials, but we have saved lives," pitched in Sarah McClanahan from Thomas Memorial Hospital. "How long is it going to take to get something back from the FDA that says we can continue saving people's lives?"

"Can the FDA approve FMT for recurrent *C. diff* with the evidence that we have?" questioned Jenny Sauk, from Massachusetts General Hospital, joining others who were pressing for an interim solution.

The discussion went way over the meeting's scheduled end time. "It looked like the doctors reached a consensus that they weren't leaving until they had a better solution from the FDA," Catherine says.

At some point, the electricity in the auditorium shut off, but the participants wouldn't leave. Instead, they lined up to introduce themselves to Catherine in the darkened room, lit only by the dim emergency lights on the floor. "They were thanking me for speaking and asking how they can help and asking to keep in touch."

Slater came over, too. "I just want to thank you for coming today," he said as he shook her hand. "How old are your grandchildren?"

"They are two years and six months," she answered, and then added, "Please take into consideration what I said."

"We will," Slater replied.

He meant it.

Two months later, Colleen Kelly of Brown University, who by that time was on Catherine's foundation's board of advisers, emailed her the news: The FDA had backed down from its original "must file an IND" stand. Instead, the agency announced that it would exercise "enforcement discretion." That essentially meant that the agency would not require doctors to file the IND application to treat patients with *C. diff* who were not responding to antibiotics.

"While we believe in the Investigational New Drug (IND) process, we did not want the IND process or our requirements to delay this therapy for the really ill patients who need it," Slater explained. The agency did, however, note that allowing FMT was a temporary stopgap measure until scientists devised better solutions. "The progress of the science will determine the long-term solutions to *C. diff* infections."

Doctors could now perform the controversial procedure to save the critically ill. But the science had uncovered only the tip of the iceberg in terms of what fecal transplants could do. The explosion of microbiome science that ensued in the few years that followed proved that fecal transplants could be used to battle all kinds of health problems.

But that progress led scientists to wonder about another use for human fecal matter: If it could cure disease, could it also alert medics when a disease is brewing? In other words, if it can be used for treatment, can it also be used for diagnostics? And that opened new possibilities—as long as diagnosticians were willing to peer into poo.

CHAPTER 15

LOOKING WHERE THE SUN DOESN'T SHINE

For the entire year of 2009, Eric Alm hadn't had a bowel movement in his own home. Instead, every time he felt that he had to go, he would indeed go—to his lab at the Massachusetts Institute of Technology, where he is a professor of biological engineering. Luckily, it was a fairly short drive, at the end of which Alm would follow a procedure similar to what John Duff did when he produced his donor sample. Alm would take a white plastic hat out of a sealed sterile bag and put it over his lab's toilet. At that point, he could finally sit down and relax—but when he was finished, he had still more work to do. Using a little scooping device, he would carve out a smudge from the hat's contents and shove it into a 15-milliliter test tube, which he then placed in a freezer.

Jumping into the car every time he was ready for a bowel movement was actually the least tedious part of the project. The more tiring part was recording everything he ate and did every single day. To make sure he wouldn't forget any detail of his busy schedule, Alm carried an iPad with a tracking app. Specifically customized for the project, the app captured everything that happened in his life. He inputted what he ate for breakfast, lunch, and dinner, recorded how much he weighed, and took notes on his energy levels, mood changes, and sleep patterns. "I would record if I got up in the middle of the night or if I felt tired, or what my mood fluctuations were," he recalls. The app knew more about Alm than his family and friends did. To have more fun with the project—and to get more stool samples—Alm recruited his graduate student Lawrence David as a second study subject.

Together, Alm and David aimed to track how their daily activities changed their intestinal microbiome. After the Human Microbiome Project described the human gut as a rich ecosystem teeming with hundreds of microbial species that help us digest food and protect us from harmful germs, scientists realized they knew very little about our inner creatures. While it's reasonably easy to observe bacteria that live on our skin or in our mouths, studying our gut inhabitants is much harder. You can't just rub a swab against your colon like you do with your skin or the inside of your cheek. Instead, doctors would have to stick tubes, microscopes, and specimen collectors into people—and who would want to do that, especially if they were in good health? And so the lives, deaths, and interactions of these creatures that quite literally live where the sun doesn't shine remain a mystery. "Most of them we know nothing about," Alm says.

The only noninvasive way to track the behavior of our microbial zoo is by studying our feces. That's what the two-person team of microbiome explorers set out to do. After gathering a year's worth of samples, they planned on sequencing them to identify the DNA of all the microorganisms from the fecal matter their intestines had produced every day.

Launching this intestinal "expedition" felt like embarking on a great yearlong voyage—even if on a microscopic scale and in a *Magic School Bus* style. "It crossed my mind that I was like an explorer," David says, recalling his reaction when Alm first brought up the idea. "I have not heard of people doing it before. That's what made this project such a joy and truly exciting." Both scientists felt that gathering the data over time was important. Taking a few microbiome snapshots would show what lived in their intestines at those particular moments, but gathering data continuously over time would show how their bacterial communities reacted to dietary changes, external stressors, and other life perturbations.

The expedition's start, however, was hard, at least for David. He almost threw up as he looked into the plastic hat that held his first sample waiting to be scooped. "As soon as I started, I immediately regretted that I signed up for it," he recalls. "It smelled so bad that I didn't think I could do it. I didn't think I was going to last through the whole year."

But curiosity won over disgust, and within a few weeks, both study participants became experts at scooping poop. Their new, extended bowel routine became the norm. Entering data actually proved more annoying than sampling because it took about an hour a day and was tedious and boring.

However, halfway through the expedition, things unexpectedly got quite exciting. Both Alm and David fell ill with stomach problems. To most people, writhing with stomach cramps in the bathroom wouldn't be a good thing, but for the microbiome explorers, a few bouts of diarrhea proved to be a boon. These episodes added some data the duo hadn't counted on—especially since they fell ill not only under different circumstances, but on different continents.

David traveled to Thailand, where he succumbed to what appeared to be a classic case of traveler's diarrhea. Of course, he had planned on gathering his microbiome samples while traveling—skipping them would compromise the study—so he came prepared. He lugged along all the requisite hats, scoops, and test tubes, the last of which he transferred into a freezer as soon he filled them—just as he did at the MIT lab. He was so well stocked that despite making a few extra bathroom visits due to his upset stomach, he never ran out of hats, and none of his loose stools went to waste. When he was ready to go home, he hired a shipping company to send his Bangkok-made samples to the lab—on dry ice, accompanied by a bunch of paperwork. "It was crazy expensive," he recalls. "To ship three to five pounds cost about $1,200, plus you had to pay for the ice."

On the contrary, Alm never left the United States. He hadn't even left Massachusetts. Yet his illness was much worse than David's. His diarrhea was so severe that it left him dangerously dehydrated, and he ended up in a hospital with an intravenous needle in his arm. He got sick shortly after eating French toast, so he suspected he had ingested some sort of common foodborne pathogen. He turned out to be right. When the hospital ran his culture, they told Alm he had contracted salmonella. Like David, Alm preserved all of his "sick" samples, except for a couple of days during which he was too weak to drive to the lab.

Once the grueling year was finally over, the microbiome explorers pulled their hundreds of accumulated samples from the freezer and batch-processed them, sequencing the DNA of their intestinal bacteria. They spend the second year analyzing data and drawing graphs and charts—and then finally published their study.[1]

The charts showed that even though Alm and David lived in the same geographic location, their microbiomes comprised fairly different bacterial species. Furthermore, the species' abundances fluctuated with what the scientists ate. Consuming yogurt, which contains *Bifidobacteriales*, increased their numbers in the gut. Eating fiber-rich foods boosted *Bifidobacteria, Roseburia*, and *Eubacterium rectale* species, which feed on fiber. The most impactful changes in the bacterial population were caused by food, Alm says—and the most impactful type of food was fiber. "The more fiber you ate yesterday, the more species you would see today," he reveals, adding that the microbes always took one day to react. "It was never the same day or two days later. The changes you saw today always correlated with the previous day. That was the pattern we picked up."

Overall, however, the species composition in the researchers' respective guts stayed fairly stable from day to day and from week to week—except when they got sick. When that happened, their guts reacted and recovered differently.

During David's Thailand visit and intestinal disturbances, the number of his *Bacteroidetes* relative to the number of his *Firmicutes* doubled, as the graphs showed. However, the changes reversed themselves two weeks after he came home, so essentially, his microbial menagerie went back to normal.

Alm had a different outcome. His salmonella encounter caused many of his bacterial species to dwindle permanently. Unlike David's *Firmicutes*, they didn't come back, but they were replaced by similar species, the graphs showed. The DNA data also clearly pinpointed when Alm contracted salmonella and when it finally disappeared from his gut. Interestingly, the new species that settled his intestine after the salmonella was gone didn't seem to make much difference in his overall well-being. This unanticipated exposure of his microbiome to new environments and pathogens might have actually shown that overall, human microbiomes can be fairly resilient.

Alm and David didn't plan on getting sick—it just happened to be an "unexpected bonus" for their study. But in the past, some scientists have actually inoculated themselves with pathogens to find out whether their inner menagerie would fight off the bugs—or not. Back then, medics didn't yet use the term "microbiome." Instead, they referred to the creatures as "intestinal flora"—perhaps a more poetic description for the diverse and lavish ecosystem that inhabits all of us. And the pathogen they experimented with was much deadlier than salmonella.

It was cholera.

Microbiome Science in the Time of Cholera

During an 1892 outbreak of cholera in France, Ilya Metchnikoff, a Russian immunologist and expatriate working at the Pasteur Institute in Paris, decided to mix some of the pathogen into his drink. Metchnikoff, who would later win a Nobel Prize for his immunology work, was eccentric, but not crazy—he drank cholera to understand how his intestinal flora would stamp out the pathogen's spread.

Medics had noticed that the pathogen responsible for the epidemic, *Vibrio cholerae*, had a peculiar modus operandi. Some people died from it, some managed to recover, and others never fell ill, despite taking care of sick family members and sharing living quarters and kitchen utensils with them. These people just seemed to be immune to the bug.

Understanding how such immunity develops could help create a vaccine, which is what Metchnikoff hoped to do. He didn't get sick after drinking his tainted concoction, so he offered it to a volunteer at his lab. The volunteer didn't contract cholera either, so another one also offered to try it. The second man, however, was less lucky. Not only did he succumb to the germ, but he nearly died. Horrified, Metchnikoff stopped feeding cholera to humans and began playing with it in a petri dish. That strategy proved productive. Metchnikoff found that some microbes fueled *V. cholerae*'s growth, while others stunted it. So he theorized that a similar bacterial battle takes place in the human gut. Therefore, the intestinal flora could help prevent disease—or exacerbate it. That meant medics could treat people by giving them the beneficial bacteria. If swallowing bad microbes made you sick, then swallowing good ones would make you healthier, Metchnikoff reasoned.

At the time, this gut hypothesis was a rather gutsy idea. Fueled by Leeuwenhoek's microscope discovery, nineteenth-century medicine viewed microbes as infectious agents rather than beneficial creatures. Researchers at the Pasteur Institute had observed how bacteria spoil food, making it putrid—and called the process bacterial putrefaction. So when medics found that microbes populate the human large intestine, they assumed that the same nasty putrefaction process was happening there, too. After all, what came out of the intestines smelled just as bad as rotten food, or worse. Scientists knew that putrid food could sicken and kill, so they postulated that the bacterial putrefaction products brewing inside the colon could do the same. French physician Charles Bouchard called the colon "a receptacle and laboratory of poisons." He argued that these poisons were absorbed by the body, continuously weakening us and making us susceptible to all kinds of diseases.

When physician Thomas Oliver translated Bouchard's work into English, he wrote in the preface that "man is constantly standing, as it were, on the brink of a precipice; he is continually on the threshold of disease. Every moment of his life he runs the risk of being overpowered by poison generated within his system." And so the medical world embraced the so-called autointoxication theory—poisoning of the body by its own fecal matter. By 1905, intestinal toxins were blamed for a vast variety of conditions—from digestive disorders to depression, melancholia, and other mental illnesses. And our large intestine was viewed as a cesspool filled with filth, rather than a vital ecosystem hosting symbiotic organisms.

Why would humans evolve to have such as a detrimental organ? As medics pondered this question further, they decided that the large intestine was a visceral leftover from the time our ancestors roamed the prairies and savannahs, where they had to run from predators and simply didn't have the luxury to stop and empty their bowels when they felt like it. So scientists decreed that humans evolved to carry their poop, along with the toxic bacterial by-products, in order to survive. But since people no longer had to run away from lions and tigers, this atavism could be corrected.

British surgeon Sir William Arbuthnot Lane advocated curing digestive disorders by removing parts of the large intestine, and even the entire intestine altogether—a colectomy. (This may sound barbaric, but modern ways of battling Crohn's disease include cutting out the inflamed parts of the intestine; unfortunately, the inflammation tends to return and afflict new areas.) From constipation to insomnia and from dyspepsia to neurosis, Lane proclaimed that colectomy was a cure for all. And patients believed him. In the early nineteenth century, he became a go-to doctor for the modish surgical cure, despite the fact that the mortality rate from the procedure was 16 percent. Those who lived, however, weren't necessarily cured. After the surgery, many patients developed worse problems than what they had before. Nonetheless, some returned for repeated surgeries. The idea of intestinal putrefaction was so convincing that Lane rid many sufferers of their large intestines before the fad faded.

These radical theories evolved partly because at the time, science had little means to determine that bacteria actually help us metabolize our meals, and partly because the functions of microbes in nature are very diverse. In many cases, the same microbes that spoil food on the outside of a body would indeed break it down on the inside—in the gut.

The roots of our symbiotic relationships with microbes date back millions of years. Bacterial life is thought to have originated in hydrothermal vents deep in the sea, where hot fluids and ocean waters bring together essential elements of life—carbon, nitrogen, hydrogen, oxygen, and sulfur. Millions of years ago, this mix may have bred life's building blocks, which eventually led to the creation of some very simple microorganisms.

At the time, our planet had little to no oxygen. The earth's atmosphere was filled with gases humans can't live on—mostly carbon

dioxide, plus some methane and sulfur, produced by volcanic erup-
tions and other geological and geothermal processes. The bacteria
that evolved to survive in that hellish place were anaerobes—the
same oxygen-hating types that happily dwell in the digesters of the
Newtown Creek, DC Water, Loowatt, and other sewage treatment
plants. However, some microorganisms evolved to produce oxygen as a by-product of their metabolism—their waste. For example,
cyanobacteria, which are thought to have appeared about 2.48 billion
years ago, generate oxygen by photosynthesis, as plants do.

At first, most of this oxygen was absorbed in various chemical
reactions—such as oxidation of metals. However, these bacteria
eventually produced so much oxygen that it began to accumulate.
Scientists call that pivotal point in the earth's history the great oxy-
genation event because it created the earth's current atmosphere,
paving the way for aerobic organisms to develop and take hold. Essentially, it created life as we know it.

But that new atmosphere was toxic to the anaerobic organisms
that liked their methane and sulfur. The anaerobes began dying, except for those that moved into the soil, water, and the anoxic insides
of larger, more complex creatures. From an anaerobe's perspective,
living in an animal's gut was just as good as living in dirt. And so
these oxygen-hating bacterial refugees moved into the intestines of
oxygen-loving organisms—and stayed there for good. From there on,
they evolved together with their hosts, moving from dinosaurs into
mammoths and from Neanderthals into modern humans. Without
them, neither our digestive processes nor our immune systems can
work. And as strange as it may sound, the functions they perform
inside our intestines are similar to what they do in anaerobic digesters. They break down larger chemical compounds into smaller ones,
releasing methane, hydrogen, carbon dioxide, and sulfur, which they
like to surround themselves with. Some of these gases come out of us
in a rather inelegant form.

We coevolved with the microbial strains that "settled" us. They
became our symbionts. However, other species that went to live in
water and soil, but didn't populate humans, didn't form symbiotic
relationships with us. When we encounter them—for example, by

accidentally ingesting them—we may become sick. Luckily, our symbionts protect us. They are territorial when it comes to foreign organisms. They produce and release antibacterial compounds—their own, self-made antibiotics—aimed at killing unwanted microorganisms—those hitching a ride on contaminated lettuce, in untreated water, or in Alm's case, on a piece of French toast. The more diverse and bountiful our anaerobic army is, the better it fights off such invaders. The people who drank cholera and didn't get sick were the lucky owners of a robust and feisty microbiome—even if they viewed their guts as toxic cesspools.

Metchnikoff believed in the autointoxication theory, but he wanted to fix the gut microbial balance without cutting out vital body parts. So he embarked on a quest for beneficial bacteria. Having done food-preserving experiments at the Pasteur Institute, he knew that lactic acid prevented milk from spoiling, turning it into a yogurt-like product. That led him to wonder if lactic acid would accomplish a similar task in the gut. "As lactic fermentation serves so well to arrest putrefaction in general, why should it not be used for the same purpose within the digestive tube?" he argued.

After studying various bacteria that made yogurts, Metchnikoff zeroed in on Bulgarian bacilli. He advocated eating sour milk products to fight off pathogens, but he also suggested that the culture could be taken in a form of a pill—essentially proposing the use of microbiome-enhancing probiotics before either word was invented. Decades before the concepts of "probiotics" and "microbiome" were described, Metchnikoff predicted their existence after the erratic results of his *V. cholerae*-drinking experiments.

Moreover, Metchnikoff essentially described what we now call "leaky gut syndrome," in which microbes escape through weakened or damaged intestinal walls and attack tissues in the body, causing chronic inflammation and autoimmune diseases. Metchnikoff didn't use the words "leaky gut," but he linked intestinal microbes, immunity, and premature aging. Earlier in his career, he had discovered phagocytes, cells that protect the body by consuming harmful agents, including bacteria or foreign particles. On the basis of known blood test results, he postulated that gut microbes can pass through the intestinal walls and inflame other organs; the phagocyte cells would then devour the inflamed tissues, thus speeding up the aging process. "The ill-health which follows retention of faecal matter is certainly due to the action of some of the microbes of the gut," Metchnikoff wrote in *The Prolongation of Life: Optimistic Studies.*[2] "The intestinal microbes or their poisons may reach the system generally and bring harm to it."

For a long time, this bacterial translocation was thought to be impossible, but studies done in the 1970s and 1980s found that gut microbes can indeed leach into the lymphatic system and the bloodstream—and make their way to other organs. Today, modern hypotheses suggest that diabetes, rheumatoid arthritis, lupus, multiple sclerosis, and Parkinson's disease may indeed originate from "misplaced" microbes that wind up in the wrong body parts. "A lot of the things he did were very prescient," says Siamon Gordon, professor emeritus of cellular pathology at the University of Oxford, who studied Metchnikoff's work. He had an almost uncanny ability to connect the dots where no one else would see any dots at all.

Having lost his first wife to tuberculosis, Metchnikoff spent his entire career on finding ways to battle diseases, develop vaccines, and extend longevity. Convinced that he had finally hit upon the answer, he touted sour milk as life-saving and life-extending medicine that fixed humans from the inside out. "The intestinal flora is the principal cause of the too short duration of our life," he stated. "Science must now set to work to correct it."

For a while, Metchnikoff's ideas flourished. In the United States, Lederle Laboratories manufactured acidophilus, another beneficial type of bacteria. In Europe, an entrepreneur used Metchnikoff's approach to launch a yogurt business, which later became the well-known food company Dannon. Unfortunately, once antibiotics were invented, Metchnikoff's ideas of balancing intestinal flora with beneficial microbes fell by the wayside. Killing pathogens was quicker and more effective than trying to rebuild the elusive and little-understood microbial balance inside the body, where neither the sun nor lab lights could reach. In the 1930s, Lederle Laboratories stopped making acidophilus and switched to antibiotics. It wasn't until the recent advances in human microbiome and DNA sequencing that Metch-nikoff's ideas returned to the scientific forefront. Now acidophilus is sold in nearly every drugstore, and medics are once again peeking into patients' insides to find out what ails them. Moreover, scientists have realized that analyzing the microbiome of a city can help govern the health of its population.

The Sewage Sleuths That Boldly Go Where No Bots Have Gone Before

It is a sunny morning in Cambridge, Massachusetts, and two young women, Mariana Matus and Newsha Ghaeli, are fussing over a man-hole cover. As they lift it, they also bring out a long, transparent plastic tube filled with wires, batteries, and electronic chips. They wince slightly at the smell emanating from the open sewer and begin to lower the plastic tube, attached to a fishing wire, into it.

The plastic tube is a robot affectionately named Luigi, after one of the plumbers in Nintendo's *Super Mario Bros.* game. Luigi is a sewage sampling robot—it collects data in yet another place where the sun doesn't shine. Just like its namesake in the game, Luigi has a "brother" named Mario—the team's earlier version of a sewer explorer. Like their Nintendo counterparts, Mario is heavyset, boxy, and weighs about 18 pounds, while Luigi is skinny, nimble, and about three-quarters lighter. The bots' creators surely have a sense of humor.

Once Luigi plops into the sewage stream, it starts taking waste-water samples. Equipped with a pump, it will eventually fill up the 250-milliliter bottle it carries. Mario used to draw samples with a needle, but that proved less efficient: the pressure squashed some bacterial cells and killed them. For all the dirty work the scientists had to do, they wanted their study subjects alive, not dead. "Luigi's pump creates a soft, gentle motion, rather than the vacuum that syringes create," explains Matus. "The samples are much better preserved that way."

Luigi can hang out in that grimy underground river for hours. When the women finally pull it up, its bottle is full of liquid brownish muck—a mix of everything that has been recently flushed or poured down the drain in the surrounding Cambridge neighborhood. Wet and stinky, Luigi is also covered with clumps of toilet paper. The latter interferes with data collection because it clogs up Luigi's machinery, so the team has spent a lot of time tweaking the bot to make it toilet paper–proof. The post-dive cleanup procedure is fairly easy. Luigi pumps some bleach through its insides, gets more bleach sprayed on its plastic shell, and it's fully disinfected. Decontaminating Mario took longer—it had to receive an ultraviolet light bath.

Matus, a computational biologist, started working on wastewater epidemiology in 2013, when she was a PhD student in Alm's lab. In 2014, Alm and urban architect and engineer Carlo Ratti launched a project called Underworlds, which aimed to do real-time sewage data collection to inform municipalities of their residents' health, and Matus jumped on the bandwagon. "I didn't really have a background in engineering or pubic health," she recalls—yet, as the team grew, other collaborators joined, including a chemist, a mechanic, an engineer, and architect Newsha Ghaeli. "As we started working together, we realized what an amazing opportunity it was," Matus says, "because we collaborated across different disciplines."

Eventually, Matus and Ghaeli spun the project off into its own company, called Biobot Analytics.

Exactly what is the team hoping to get from the ghastly bottle Luigi brings up from the underbelly of Cambridge? The sewage is a trove of data, says Matus. You can discover a lot about a city by passing its sewage through a DNA sequencer. You can learn about its residents' health. You can find out what medications they take. And you can catch brewing infections before they reach epidemic proportions.

Like blood, human stool carries a lot of information about the body. It contains detectable traces of metabolites, hormones, antibiotics, stress markers, depression meds, digestive aids, allergy drugs, and painkillers, among other things. And while sewage sampling can't give you health information for any particular "contributor" to that wastewater—and thus preserves residents' privacy—it can give you a broad view of what's happening in a given geographic area. One day, multiple Luigis may be roaming the vast sewage labyrinths beneath your feet, detecting pathogens in real time and sending instant text alerts to health officials. In other words, sewage sleuths are made to float deep underground, but the sky may be the limit to what they can do.

Biobots may be able to identify the specific strain of influenza going around—and inform doctors that they need to vaccinate residents for exactly that strain. They can detect a spike in populations of *E. coli*, norovirus, or other bugs, and alert medics before an outbreak happens. If some residents contract a bacterial infection that becomes resistant to antibiotics, they may start shedding this mutated bug into the sewage. The mutant strain could pass its antibiotic-resistant genes to other microorganisms—and the bots can detect this brewing public health hazard before it wreaks havoc on society.

Epidemiologists may solve these pathogen problems with chemicals, but Andrea Silverman, who holds joint appointments at NYU's Department of Urban Engineering and Department of Global Public lic Health, had been experimenting with another treatment—sun. When wastewater is exposed to sunlight, it creates reactive oxygen molecules, which do exactly what their name implies: they react with microorganisms, breaking their cells. In one of her projects, Silverman experimented with an open-water sewage wetland, whose content is naturally disinfected by the sun. There are certain chemicals that can make the photochemical disinfectant process more efficient, Silverman says, but not every city or municipality has space to build wetlands.

Sewage: The Final Frontier

Luigi and Mario may be the first biobots that sift through sewage for useful health information, but humans have done it before. Some are doing it on a regular basis as part of epidemiological surveillance.

Every week or two, teams of WHO health workers in New Delhi, India, sample the city's sewage. Only they do it manually, without the help of sophisticated robotic machinery. They don gloves, masks, protective coveralls, and high rubber boots—and bravely venture out into the sewage stream with a small bucket. In some places, they can toss the bucket in, but in others, they have to wade to midstream to get the "representative sample."

When the bucket comes back, they filter the wastewater through a mesh to sift out the typical sewage components, then pour what's left into bottles destined for testing labs. The bottles travel to the lab in iced containers because their contents are truly precious: they reveal the country's current state of health. When the lab reports come back about two weeks later, they list all the typical sewage microorganisms—which is fine, says Lucky Sangal, India's national professional officer for vaccine-preventable disease, who oversees the operation in New Delhi. "What we don't want to see is polio. Nowadays we don't see it. And that's good."

Poliomyelitis, a crippling viral disease that attacks the spinal cord, leaving some children paralyzed, used to be endemic in the country. "In the early to mid-1980s, the entire world used to have 1,000 cases of polio every day, and India used to contribute 500 of them," says Deepak Kapur, chair of India's National PolioPlus Committee for Rotary International, a nonprofit that takes part in the Global Polio Eradication Initiative (GPEI). Thanks to India's continuous immunization campaigns, the disease has been eradicated in the country—it has been officially polio-free since March 2014. But that doesn't mean the country's health officials can let their guard down. The virus is still present in neighboring countries—Pakistan and Afghanistan—so one never knows when it may cross the border inside an infected traveler. Polio is tricky. It can leave a person completely disabled, or cause no symptoms at all. So someone can be harboring the virus and not know it. In the meantime, the virus can be replicating inside the person's intestines, escaping with fecal matter.

That's why Sangal's team dips their buckets into New Delhi's murky wastewaters. Their colleagues do the same across the country in 52 designated sampling spots, chosen because of their locations. "They are usually set up at the catchments that drain sewage from urban slums—the population mixing bowls," Sangal explains. If a sample tests positive for polio, heath authorities immediately vaccinate all children under five that live in the area.

Moreover, if a sample tests positive, scientists can now tell where the virus came from. As viruses procreate and evolve, their genomes change slightly. Every time a poliovirus sample is collected from a person in a particular geographic location, the medics enter its exact genetic characteristics into a global database. Over the three decades of eradication efforts, health workers have amassed a huge amount of information about the virus's genetics and built a "polio family tree." They can genetically match a viral strain from a city's sewage to that family tree and pinpoint its "birthplace," Sangal says. "It's next to impossible to identify the person who is shedding the virus, but based on the molecular genetics of the virus, we can tell what province or country it came from," she says. "And then we can alert the health authorities to vaccinate children there."

Across the border in Pakistan, Sangal's counterpart, Nosheen Safdar, oversees similar collection efforts in Islamabad, while her colleagues do the same at Pakistan's other collection sites—60 in total. Nigeria, which hadn't seen a polio case for three years now, had 113 as of early 2020. "We can't say 'we are polio-free,' if we are not actively looking for it," says Tunji Funsho, chair of Rotary International's Nigeria National PolioPlus Committee. And that means sifting through sewage for polio to make sure there is none.

Stool, the Diagnostic Tool

Just as sewage content reflects a city's overall health, the microbiome is a good indicator of an individual's well-being. That's why Alm thinks that both types of tests are essential for understanding societies' health. In the future, these data might become a part of a daily health check routine for both cities and the individuals who live in them. In this scenario, conducting sewage health checks would be fairly straightforward: just use biobots. But personal tests would be trickier. Even though several companies already offer sample collection kits for those who want to sequence their gut microbiome, most people wouldn't want to scoop their poop and send it to a lab on a daily or even weekly basis.

Alm thinks the answer is in smart toilets. In his vision, these futuristic devices would be outfitted with scoops and sensors that would grasp a sample from your morning bathroom visit, run it through various reagents, and send it to the lab. Such smart toilets would produce your daily metabolite and microbiome report, perhaps even with a recommendation to eat more fiber today and stay away from sugar after yesterday's bingeing. Or perhaps your toilet would tell you that it detected a trace of salmonella in your gut and you should quickly pop a few probiotic pills. Better yet, your toilet might log your daily report into a well-being database, which your personal physician would glance through during your next visit to see whether you experienced any health blips.

This scenario may sound overly futuristic, but it actually wouldn't be that hard to implement technologically. The latest toilet models have come a long way since Sir John Harington's Ajax of 1596. They are already so smart they're nothing short of geniuses. The well-known Japanese manufacturer Toto makes toilets that minimize water use by spraying the bowl with mini-streams in a "tornado flush" fashion and prevent bacterial buildup by using electrolyzed water, made by a pair of built-in electrodes. Besides cleansing themselves, these toilets can also clean the people who use them by shooting water and then blowing hot air on the appropriate areas—which eliminates toilet paper use and can be managed with a remote control. In Japan, this handless cleansing service has been a standard feature of toilets for decades, but Americans have never embraced it to the same level, even though Kohler makes models with similar features. What's more, the latest johns are Bluetooth-ready, can connect to the internet, and can play streamed music—which means they're already capable of texting with your doctor.

The issue isn't the technology itself, but the usage logistics. Once toilets begin to test you, they stop being simple home appliances and become medical devices. If they need repairs, a call to your average neighborhood plumber will no longer suffice. Dealing with their sophisticated machinery, chemical reagents, and sensors will require someone with at least one advanced degree. And the company manufacturing these smart diagnostic modules will have to assume responsibility for all the errors, stresses, and expenses their inadvertent malfunctions could cause. Imagine getting a morning message from your john about an advanced-stage cancer biomarker, which wreaks havoc in your life, leads to a slew of costly and painful tests that find nothing—and eventually proves to be a silly sensor glitch that causes the company to issue a recall. That really is the hurdle that toilet diagnostics would have to overcome to become commonplace.

And yet these smart toilets are already being developed. For now, they reside only inside research labs. For example, researchers at Canary Center at Stanford are developing a smart toilet that analyzes urine, looking for common problems like diabetes and kidney stones. But the team aims to expand into methods for detection of other conditions, including parsing stool for specific molecules that can warn them about developing disease before it manifests itself.

Figuring out how to navigate this smart diagnostic future is a challenge. But Alm thinks it's not only possible, but simply a matter of time. "Never bet against progress," he jokes—because progress is inevitable.

In fact, just in the past five years, we have indeed made progress. Today's *C. diff* sufferers don't have to give themselves homemade enemas in their bedrooms. They can find a medical doctor who will do the procedure in a clinic using a carefully screened sample ordered from an authorized provider. Like the Buddhist monks who buried clay jars underground, modern medicine has learned to stock up on its own version of curative muck. Only today we fill freezers with bottles to keep these therapeutics ready for those who need them when they need them.

Sounds improbable? These institutions are called stool banks. And multimillion-dollar corporations are arguing over the rights to all this shit.

CHAPTER 16

FROM THE KINDNESS OF ONE'S GUT: AN INSIDER LOOK INTO STOOL BANKS

A petite woman clad in a lab coat, face mask, and hairnet sits at a biosafety cabinet, gingerly dipping a pipette into a large plastic bag filled with dark, viscous liquid. Designed for making medicines, a biosafety cabinet is somewhat similar to a lab bench, except for the drop-down glass screen separating the worker from the substances she works with. Other than the woman's double-gloved hands that slide underneath the screen, nothing is allowed inside that sterile environment—even the air inside must pass through several filters to eliminate all particles.

In front of the woman is an encapsulator, a square platter with several rows of capsules—or rather, half capsules, with their open ends facing up, waiting to be filled. Her pipette isn't the typical one-tuber you may have used in your chemistry labs, but rather a "five-nozzle" one, with five openings in the bottom, all of which draw and discharge simultaneously to fill five capsules at a time.

As the woman releases a button on the pipette's top, the dark liquid from the plastic bag starts climbing up the nozzles. Once they are about three-quarters full, she carefully moves the pipette over to the encapsulator and positions the nozzles over the first five capsules, releasing the content into them. Within a few seconds, the dark, gooey drops fill up the capsules with one of the most potent medicines ever discovered by humans—their own fecal matter.

Intended for oral consumption, the capsules are made in the lab at OpenBiome, a nonprofit organization that operates a stool bank that gathers and preserves samples from healthy donors and supplies them to doctors for *C. diff* treatments and research. Located in a suburb of Boston, not far from the Massachusetts Institute of Technology, OpenBiome is known far beyond the state and even the country. Researchers and physicians worldwide have used its services for various reasons. I find this fact both ironic and existential. For decades, doctors and health care workers worldwide had been striving to eliminate the fecal-oral contamination route—and then began to realize how therapeutic this route could be, if used properly. Launched in 2013, the bank gathers stool donations every day. When they arrive at the lab, they are processed quickly, according to a strict protocol, says Delasie Dela-Seshie, the manager of the OpenBiome manufacturing program.

The woman dips her pipette into the bag again, filling the next quintet of capsules—and then another one, as perfectly as she has done the rest. The fecal matter must end up exactly inside the capsules, without even the smallest droplet escaping. She does it again and again, each time with the same reverent precision, like a jeweler fitting a precious stone into a delicate silver frame. I watch her moves, mesmerized.

"She's amazing," I say to Dela-Seshie under my breath. "Her arm doesn't twitch even one single bit. Doesn't she get tired?"

My question draws a happy smile from Dela-Seshie's face. "That's my team," he says proudly. "They *are* amazing." Then he explains further. "You have to be qualified on the process, so everyone who does it has mastered the task. The operator who is doing it here probably has done it thousands of times."

When the woman is done, she places the encapsulator's top panel of empty half capsules onto the filled ones, and presses hard. Inside, the halves fuse together into medicinal pills. Before finishing, the woman briefly examines them one by one to make sure that the capsules—christened "crapsules" by some—are tightly closed and ready for consumption. After that last check, the pills will be transferred into a storage room full of freezers holding thousands of samples waiting to be used. Until then, they are kept frozen at −80°C. At that temperature, the microbial zoo goes to sleep, meaning that it stops eating, breathing, multiplying, and dying, so the samples remain in their pure, original, unadulterated form. Once patients swallow the capsules, the microbes wake up and begin to colonize their new home. To restore their gut microbiome, patients must swallow a handful of them rather quickly—30 pills within 90 minutes.

The crapsules aren't the only way OpenBiome's medicine can be administered. The company also makes bottles used for "lower delivery"—the same method Catherine Duff and her husband used, except that they are meant to be used in medical settings. OpenBiome does not ship to individuals, only to qualified medics and institutions. Another form of fecal matter transplantation, or FMT, is via a nasal tube that runs all the way into the stomach. These transplants require a smaller dose, so OpenBiome produces them in vials. Generally, OpenBiome likes to have every donor's samples present in all three transplant forms, Dela-Seshie says. That helps ensure sample diversity. So once the woman I am watching is done with the capsules, she will fill some vials the nasal delivery, and some bottles for the lower-delivery method, too.

Preparing donated fecal matter is a multi-step process. The samples arrive at the lab in tightly closed plastic bowls—although Dela-Seshie suggests a more correct term: "We call them 'specimen collection containers.'" Before they can be opened, the crew preps the biosafety cabinets, wiping them down thoroughly, first with various cleaning agents and then with alcohol to remove the cleaning agents' residue. No microbes and no air particles are allowed to contaminate them. The workers wear gloves, masks, and hairnets not so much to protect themselves from the fecal matter they handle, but to make sure they don't accidently contaminate the future transplants, Dela-Seshie tells me. That was the intent of the sterilization protocol John Duff followed, too.

Once the cabinets are properly sterilized, the plastic bowl can be opened and examined to see if the sample meets OpenBiome's standards, Dela-Seshie says. Stool donations must have certain physical characteristics. They must be of a certain consistency—a perfect combination of solid and soft. If a sample is too hard, it will be too difficult to work with. A sample that's too liquidly may be a sign of a health issue manifesting itself in slight diarrhea, so the donation will be ditched.

How do you decide when your sample hits that sweet spot? I inquire. That's what the Bristol Stool Chart is for, Dela-Seshie tells me, pointing to a colorful poster on the computer screen. The chart depicts seven different textures and shapes of stool, along with vivid descriptive terms, including "lumpy," "mushy," "smooth like a snake," and "with cracks on the surface." "We only accept the samples that fall within categories three to five," Dela-Seshie says. It almost sounds like a stool beauty pageant.

Samples that pass the beauty test will be "groomed" into therapeutics. They will be transformed into the smooth, easily flowing milkshake form that John Duff created with his kitchen blender. Another worker, a tall, slender man in gloves and a mask, demonstrates the process. First, he mixes the stool with liquid similar to the enema solution Duff used. Then he adds a cryopreservant—something that Duff didn't need, but OpenBiome does because it stores live microbes in a frozen form. The microbial cells are full of water, and the cryopreservant prevents this water from freezing and bursting the cells, Dela-Seshie explains. It's harmless to humans and has no effect on the microbes' well-being or therapeutic qualities, he adds. It's even safe to use in foods.

Meanwhile, the worker moves on to the next step. Holding a large plastic bag with a fine inner mesh that separates it into two halves, he loads the entire sample into one half. He seals the bag and hangs it inside a homogenizer—a refrigerator-sized apparatus with two large paddles on the sides. For the next few minutes, the paddles smash the hanging bag like partying children pummeling a piñata. With every smack, the little lumps of stool blend into mush, which slowly seeps through the bag's mesh from one side to the other. The paddles and the mesh essentially function like an industrial-sized blender and cheesecloth, sparing the operators some of the unsavory manual work DIY stool donors had to do.

As I watch the paddles whip the sample into a transplantable milkshake, Emily Langner, OpenBiome's director of external affairs, tells me that this demo is very popular with children. "We sometimes show this at science fairs for kids," she reveals. "They think it's awesome. And yeah, we use ice cream instead of poop."

The paddles stop, and the operator brings the bag back to the biosafety cabinet to examine the result. He opens the bag and looks satisfied. "He can visually see that the sample is homogenized," Dela-Seshie comments as the man begins to draw the dark emulsion into his pipette, filling up little bottles labeled with OpenBiome's green logo. Once he is done with the bottles, vials, and crapsules, another plastic bowl will probably be waiting for him already. OpenBiome receives up to 40 donations a day. Because most of its procedures are manual, this can be a lot to process for a team of about six people. "It can be tedious, but knowing what it's for makes the work easier," Dela-Seshie says. "That's the mentality of the team."

That was the mentality OpenBiome was originally founded on. It was also how the stool bank got its name: the founders thought that every patient deserved open access to a fecal matter transplant.

Two Thanksgivings and a Love Story

OpenBiome was conceived at two Thanksgiving dinners, thanks to a love story and a case of *C. diff.*

In 2012, Mark Smith, who was working on his PhD at MIT, and Carolyn Edelstein, a student at Princeton University, were dating. Edelstein's grandparents lived in the Buffalo area of New York State, and that's where her family typically congregated for Thanksgiving. This year, she invited Smith to join her, and he did.

Little did they know that this Thanksgiving gathering would have consequences stretching long past the upcoming holiday season. It would shape their relationship, change their career paths, and completely transform their lives in a way they couldn't imagine. It would also make the terms "poop" and "stool" some of the most commonly used words of their everyday speech. Moreover, the ripples caused by that gathering would echo far and wide, changing how doctors study the microbiome's role in health and disease all over the world.

Even more interestingly, neither Smith nor Edelstein was a physician, let alone a gastroenterologist. Edelstein was studying public policy and economics at the Woodrow Wilson School of Public Policy at Princeton and worked for the United States Agency for International Development (USAID). Her USAID office focused on developing innovative ideas aimed at improving health and well-being on the international level. And Smith was a computational microbiologist, sequencing bacteria for his research. But family gatherings sometimes work in mysterious ways.

At that Thanksgiving dinner, Smith met Edelstein's cousin, who had traveled from New York City despite having been very sick for several months. Once a young, healthy twenty-something college graduate, he had gone to a hospital for a routine surgery—and had come home with a *C. diff* infection. His story followed the same trajectory as Catherine Duff's: repeated rounds of vancomycin that yielded little help. A bit luckier than Duff, he had been able to keep working through his recurring infections, but he was withering from exhaustion and gloom. While he refused to let the infection to take over his life, it slowly robbed him of his health and happiness. An ambitious, career-oriented individual, he had seen his existence reduced to going to the office and persevering through the day. And he knew every public bathroom on the way home.

Having been a pre-med student at Harvard and working in health care finance, he knew more about medicine than many other *C. diff* sufferers. He had done his own research, he had learned about fecal transplants, and he had even found a doctor in New York who administered them. The problem was that the one and only fecal transplant provider in the city had a six-month waiting list, and Edelstein's cousin didn't want to live in misery for that long. He was considering doing the transplant himself, and he mentioned this radical idea to his family.

C. diff and diarrhea were hardly Thanksgiving table topics, but with Smith as a guest, they found fertile soil. "Oh my goodness, you are a microbiologist?" Edelstein's cousin asked her boyfriend. "Can you tell me what I should do?"

Most microbiologists would probably be clueless about the topic. But Smith was quite familiar with it from his bacteria sequencing experiments. At the time, it was a hot new thing. Scientists were sequencing bacterial communities from people or mice and pondering how those microbiomes affect health and disease. But they were facing a chicken-and-egg problem—they couldn't tell which came first. It was impossible to tell whether a particular bacterial combo caused disease or resulted from it. So Smith was looking into ways to alter microbial communities to see which interventions proved curative and which made things worse. Transplanting feces—in mice, for example—could be one way of doing so.

"Sequencing bacteria was cool, but intellectually unsatisfying because you don't know if microbes cause the disease or respond to it," Smith says, recalling his research quandaries at the time. "A fecal transplant was a cool way to manipulate the system. You could change the microbial community, establish causality, and develop correlations, which is what the field needed." For Smith, FMT was a topic of academic interest, but now it appeared to have a more important application.

The two men totally hit it off. They talked about fecal matters through the evening. They cited medical literature they had read. They discussed FMT's effectiveness. "My cousin was seriously considering doing it and he wanted to make sure he wasn't crazy," Edelstein says. "So he just kept talking to Mark."

As the guests parted, they left with newly formed resolutions. Edelstein's cousin decided to try a DIY transplant. And Smith and Edelstein decided that patients like him should not have to wait six months for a simple and highly effective procedure. So they resolved to create a stool bank that would have samples ready when patients needed them. Their education and experience fit the goal well. Smith would use his microbiological knowledge to devise the transplant manufacturing process, and Edelstein, well versed in the nuances of policy aimed at improving human lives, would work out the quirks with the FDA.

When Smith got back to Boston, he started talking to local doctors to understand what the barriers to providing transplants would be. The challenge, he found, was in properly screening stool donors and producing usable transplants in a scalable way. "The idea of starting a stool bank seemed an obvious solution to the problem, so if nobody was going to build it, we would build it ourselves" Smith says. "We just needed to figure out an operational solution."

Figuring out such a solution took time and creativity. The one-off process used by desperate *C. diff* sufferers wouldn't work. "Everything that touches the sample must be sterile and disposable," Smith explains. "People historically put stool in blenders and passed it through coffee filters, but that wouldn't scale up." It was also unsustainable. Smith needed something that didn't require throwing away a contaminated kitchen appliance every time a sample was processed. He also needed something more robust and automated than manually squeezing the "milkshake" through a cheesecloth or a coffee filter, but he couldn't find any good substitutes among the available pharmaceutical equipment. That business just wasn't set up for processing poop.

The solution came—once again, just like blenders and cheesecloth—from the food industry. To turn food ingredients into smooth emulsions, food companies use bags and homogenizing machines. That's where those piñata bags I watched being smacked in the homogenizer came from. Food producers fill these bags with fruit or other substances and beat the stuff into pulp. Smith found the bags during his operational solution quest and used them to squash his own stool into mush. "I spent many late nights in my MIT lab doing that," he recalls with a chuckle. "Those first samples weren't intended for clinical use, but just to prove that the process worked."

Luckily, no one in Smith's lab was shocked by his eccentric experiments. His academic environment was extremely conducive to this work. He was part of Eric Alm's lab, where several other stool-intensive projects were in various stages, including one for the Department of Defense, which looked into building an artificial gut on a chip. So Smith's late-night pulping fit in perfectly—and that's how homogenizers became part of the manufacturing process. "I finally got the material into a form that would be stable in long-term freezing storage," he says, recalling one of his success milestones. "And that's what we use today."

As he was figuring this out, Smith enrolled in MIT's Venture Mentoring Service, which paired up budding companies with experienced businesspeople for advice and networking. It also gave them a chance to pitch business ideas to potential investors.

At his first pitching session, Smith was to present his case to a panel of several businesspeople. Joining by phone was also a successful medical entrepreneur Smith hoped to get advice from. He was barely halfway through his speech when the phone participant suddenly interrupted him with a seeming non sequitur: "Is Ashton Kutcher in the room?"

"Ashton Kutcher?" Smith echoed, confused. He was prepared for any stool-related inquires, but not for celebrity trivia. "No. Why?"

"Because I feel that I am being punk'd!" the man responded, referring to the popular reality TV show *Punk'd*, run by Ashton Kutcher and famous for pulling pranks on Hollywood stars and other celebrities. "This is the most ridiculous thing I've ever heard," the man continued angrily. "I am a doctor and I've never heard of this fecal transplant! This is totally crazy. How dare you waste our time with this nonsense?"

That was a disheartening blow. "We were hoping to hear something along the lines of 'OK, here's what you do to start a company,'" Smith says—and instead it was a cold shower. "I followed up and sent him all the papers showing that there is real evidence behind this," he recalls—but he never got a response.

The incident set the startup's efforts back several months, but not forever. "There were other times when we presented this idea and ended up getting laughed and giggled at," Smith recalls. "But we persevered. We decided to do it anyway and let the evidence speak for itself."

In May 2013, Smith went to the FDA workshop where Duff delivered her passionate speech, after which the agency gave FMT a green light by exercising enforcement discretion. That was a boon for OpenBiome. Originally, the founders were planning to create their own IND and have doctors who needed a sample join that IND—they would essentially join in the already existing research rather than go through the hassle of creating their own. But when the FDA deemed INDs unnecessary, things became even easier. OpenBiome keeps a so-called biologic master file with the FDA. This file contains all the manufacturing components of the IND application, explaining the methods, equipment, labeling, and all sorts of other details the agency wants to know about the making of an investigational new drug. And physicians ordering an FMT dose don't need to join OpenBiome's IND. They simply need to reference the IND in their request—and the shipment goes out within hours.

Another boost came with a grant from the Neil and Anna Rasmussen Foundation that allowed OpenBiome to hire staff. "Until that point everyone was a volunteer," Smith says. "Thanks to the grant, we were finally able to hire our first employee and that was a very exciting moment." In the summer of 2013, OpenBiome produced its first batch of FMT samples. In October of that year, it sent its first therapeutic shipment to a patient. Another five shipments were soon on the way, too.

Before OpenBiome's founders knew it, the next Thanksgiving season was coming up, and Edelstein's family was reconvening in Buffalo. Needless to say, fecal matters were on the table once again. But the tone and flavor were quite different.

Edelstein's cousin was cured of *C. diff*. He did his transplant himself, in his own apartment, with a roommate's stool. His outcome mirrored other patient stories—he was back on his feet the next day, and his diarrhea was fully gone. At the same time, OpenBiome had established itself as the go-to facility for stool samples so that patients like him wouldn't have to wait. Edelstein's cousin moved on with his life—and so did OpenBiome's founders. Having sent six shipments in 2013, they aimed for a thousand the next year. "That was a very ambitious goal," Smith reflects on the number. "And we exceeded it. We served 1,400. We scaled absurdly quickly." Within the next five years, OpenBiome exceeded all growth expectations. By early 2019, the nonprofit had sent out 46,000 treatments. Moreover, researchers from across the globe began to order OpenBiome's FMT samples to study how to cure other microbiome-related diseases, from multiple sclerosis to malnutrition in children. One study, in St. Petersburg, Russia, used OpenBiome's FMT to alleviate intestinal distress caused by chemotherapy in children battling cancer— with reasonable success. Another study, at nearby Harvard Medical

School, used OpenBiome's crapsules to investigate whether changing one's microbiome could reduce peanut allergies. Participants took a load of OpenBiome pills and researchers monitored their tolerance of peanut protein over time. "The results are promising," says Rima Rashid, who is running the study. Some patients' tolerance really improved, but the study isn't finished yet. If this works, it would tremendously improve the lives of those who are so dangerously allergic to peanuts that they can't travel or eat at restaurants. "Peanut allergies are responsible for 80 percent of all fatalities in food allergies," she says. "It really affects people's social lives and causes a lot of anxieties."

This rapidly expanding research required OpenBiome to have a steady stream of suitable donors. And finding them proved quite challenging.

From the Kindness of One's Gut

An energy engineer freshly out of college, Leland Baldwin wasn't familiar with the rapidly developing field of stool therapeutics until his sister, a physician assistant, sent him a link to the OpenBiome's invitation to take a stool donor's test. "You are in the Boston area," Baldwin's sister wrote. "You should see if you are eligible."

Baldwin clicked on the link, read about the concept, and was intrigued. "The fact that I can help people by pooping was so funky and weird that I wanted to try it," he says. It seemed like he would have made a good candidate, too. He lived very close to the OpenBiome facilities, he was in good health, and he was fit and exercised regularly.

Plus, he was neither turned off by the subject nor had a problem dealing with poo, which certainly helped. "It would be harder for me to do it with somebody else's poop, but I was OK handling my own," Baldwin says. The fact that OpenBiome also offered to pay for donations—or to be precise, compensate donors for their time— didn't hurt either. It wasn't going to make anyone rich, but it made trying this unconventional endeavor more attractive.

Baldwin entered his information on the web form. He also filled out a long health questionnaire and received an invitation to undergo

FIGURE 7. A campaign poster at the OpenBiome headquarters urges people to donate their life-saving stool for transplantations. Yet joining the ranks of stool donors is harder than getting into Harvard or MIT. Only about 3 percent of applicants pass all tests. CREDIT: LINA ZELDOVICH

a full screening. The testing was exhaustive. It included several onsite visits and interviews, plus a blood and a stool sample test. The testing lasted a few weeks. And then the "acceptance letter" arrived. It thanked Baldwin for taking the time and effort to go through all the tests and congratulated him on passing all of them.

"It wasn't quite as over the top as the college acceptance," Baldwin recalls—but knowing he was in good enough health to donate stool to help cure people was exciting. "It felt good to know I could do that."

Baldwin had every right to feel good about acing the tests. Statistically, joining the ranks of stool donors is harder than getting into Harvard or MIT. "That's the joke we like to say here, but it's also true," says Majdi Osman, OpenBiome's clinical program director. "Only about 3 percent of the applicants pass all tests." Ivy League colleges are indeed easier to get into—their acceptance rates range from about 4.5 to 10 percent.

Finding healthy donors living close enough to the stool bank is a real challenge. It is amazing how many people have a health issue or harbor a disease—without even knowing it. In 2014, Australian researchers at the University of Melbourne and the University of New South Wales in Sydney did a study to assess what's involved in finding suitable donors—and came to that exact conclusion. Of their 116 study participants, recruited after seven months of hard work—via letters, newspaper ads, and online invitations—few earned their donor status at the end. Nearly half dropped out even before the testing began. Once they heard they would have to donate five times a week for a minimum of six weeks, they wanted no part of it, despite the promised monetary compensation.

Things didn't get much better after the testing began. Twenty-seven participants were found to have various health issues, including six people with risk factors for Creutzfeldt-Jakob disease—a very rare degenerative brain disorder that typically afflicts only one in a million people. Having so many study participants at risk for such a rare condition was strange in itself. But it was after the researchers narrowed their selection to what they thought would be a truly healthy group of 38 people that the participants really began dropping like flies.

Five of them harbored *Dientamoeba fragilis*, a microorganism that lives in pigs, gorillas, and some humans—and can cause diarrhea and fatigue and interfere with children's growth. Five others had *Blastocystis hominis*, a microorganism often found in contaminated water and known to cause intestinal distress. One person actually harbored both bugs, another one had norovirus, and a third may have had hepatitis C. More impressively, one participant tested positive for *Giardia intestinalis*, a parasite that invades the small intestine, along with toxins released by *C. diff* bacteria—the very plague the stool donations are primarily used to treat. Altogether, 15 out of 38 were disqualified due to various latent infections—and yet they had no symptoms.

More testing shrunk the healthy pool further. Some were excluded because they were overweight, some because they started antibiotics during testing, and others due to anxiety and depression. (The microbiome plays a role in mental health, too, so people with such conditions can't donate.) Finally, one was thrown out due to illegal drug use, and another due to legal drug use—because the drug caused constipation. At the end, out of the 116 prospective donors, only 12 stood strong—a fact that puzzled the research team. It's not uncommon for people to carry intestinal parasites without being sick, or to have elevated risks of some diseases, but not in such high numbers.

"We did not expect it in such a high proportion," Sudarshan Paramsothy, one of the study's authors, said as he presented the results at the American Gastroenterological Association's conference in Chicago. "Our screened donor population was not an at-risk group." The researchers' conclusion was that "it is difficult to identify appropriate and willing anonymous donors" for this therapy.

OpenBiome's experience echoes the study's findings. "Some people may harbor an organism and be completely asymptomatic," Osman says, which is why OpenBiome's testing is so extensive. "We screen for autoimmune conditions, allergy, asthma, diabetes, parasites, and recent antibiotics, among other things," he explains. OpenBiome clinicians go through the questionnaires, measurements, and blood counts with a fine-tooth comb. So as donors move through the sequence of tests, they get rejected for a variety of reasons. From allergies to traces of past infections to lifestyle choices, any little blips can raise a red flag. The last testing phase, which includes blood and stool samples, checks for infectious diseases such as HIV or hepatitis, viruses and parasites, and antibiotic resistant bacteria that may lurk in people unbeknown to them.

Recent travel can disqualify an applicant if it includes countries with high rates of infectious disease. But unlike people with lasting medical issues, travelers don't get barred forever—they just have to sit through a quarantine period to make sure they don't harbor infections. "Travel disqualifies you from donating temporary," Osman explains, "but diabetes would disqualify you permanently."

Even minor flukes in a body's complex biological machinery can result in a rejection. One donor recalls that she didn't get past the early stage because her BMI—Body Mass Index, indicative of how lean or overweight a person is—was out of the range allowed. "Basically, I was rejected because I was too fat," she says. "And my friend got the same response."

Moreover, joining the donor pool doesn't mean you're qualified to donate continuously. Every couple of months, donors must undergo a retest to make sure they haven't picked up a bug or developed a health issue. Additionally, the frozen samples they have already donated can be shipped to patients only after their retest comes back clear. That "quarantine" period ensures that every stool sample the donors produce is pathogen-free. If retesting indicates a problem— think *E. coli, Salmonella*, antibiotics. or a sudden sugar spike—all the frozen bottles and capsules made from that donor's material are discarded.

There's an age cutoff, too—donors must be between 18 and 50. "With age, other health issues increase, and we want our donors to be as healthy as possible," Osman says—and the 50s is the age when even very healthy individuals may start having health blips. The average donor age is around 27. Most donors are graduate students and young professionals in the Boston area.

To find volunteers, the company runs advertising campaigns with local media organizations and on college campuses. "A lot of people find out about us by the word of mouth and through news stories," Langner says. "We are lucky that there's lots of local interest." It helps with the ongoing challenge of finding the perfectly salubrious 3 percent.

That's why Baldwin had every reason to be proud of himself and his health. He had no *D. fragilis, B. hominis, G. intestinalis*, risk factors for genetic disease, or other maladies commonly afflicting the other 97 percent of people. After his acceptance, Baldwin's daily life included a new, previously nonexistent step: stopping by OpenBiome's facility for a bowel movement. When Baldwin arrived, a technician would give him a plastic bowl with a lid, plus a special frame that held the bowl in the toilet. "The frame sits under the toilet seat," Baldwin explains, "so you have to aim a little bit, but it's not hard to do." If he wanted to, he could also get a pair of gloves, but he didn't think they were necessary. "It's not like you are touching it," he says.

Once done, Baldwin would cover the bowl with the lid and write the time and his donor number on the top. He would place the container into a plastic bag, seal it, and throw away the frame. Then he would thoroughly wash his hands, leave the bathroom, and hand his donation to the technician. As he continued with his day, it would join other samples waiting to be processed by Dela-Seshie's team.

Baldwin would donate four to five times a week, which was slightly higher than average. "Most donors come three times a week, sometimes four," Osman says. "We don't obligate donors to come in, because we recognize that they are busy individuals who have jobs and lives. But most are committed and come in about 3 to 4 times weekly. They work it into their daily routine."

That's what Baldwin did. On weekdays, he would donate on his way to work or on his way home in the evening. On weekends, he'd bike to the place. He grew accustomed to the routine and actually enjoyed it because he was attracted by the science behind the therapeutic application. "My friends thought it was funny and weird, and my family thought it was cool. My parents are scientists, my sister is in the medical field, so we thought of it as cool stuff."

Among all the aspects of a stool donor's life, there was one specific feature Baldwin came to look forward to: OpenBiome's Friday emails. These weekly letters told patients' stories, chronicling their horrendous battles with *C. diff* and their miraculous recoveries thanks to OpenBiome's donor's samples. "Every week they would send out an email from someone who did their treatments," Baldwin recalls. "It was a feel good Friday story."

"Happy Friday, donors," an email like that would start. "I was never able to exist more than four days off antibiotics before the infection returned in an amplified form each time," one email read. "My illness forced us to cancel all summer vacation plans. We were even unable to participate in even normal activities such as attending shows, eating out. After the microbiome treatment, I am once again eating normally and enjoying good health. I am profoundly thankful to OpenBiome and my doctor. Your work is meaningful, significant and life-changing."

Some notes were longer and more detailed. "I cannot thank you enough," one woman wrote about her *C. diff* battle. Once a healthy, active, 71-year-old who still worked and ran two miles a day, she was left housebound, in constant pain, and unable to see her family. "My life was taken away from me. . . . Coming home Christmas day from visiting our grandchildren, my husband had to find places along the two hour route where I could use the bathroom . . . the rest of the time, I laid in a fetal position in the car holding my stomach. . . . I knew I had to do something as my son was getting married in Florida in May 2016, and there would have been no way I could have gone there with active *C. diff.*" By that time, she had received her transplant in a hospital from a gastroenterologist. With "the donor stool provided by OpenBiome . . . my life was given back to me . . . I truly don't know if you can imagine the immense gratitude I feel towards your organization, and whomever the donor was who gave my life back to me."

It felt good helping people, Baldwin recalls. "For me, it was just a kind thing to do."

OpenBiome doesn't pay its donors per se, but it compensates $40 per donation for their time commitment. Osman says it's only fair. "Our donors commit a lot of time just to be a part of the program. They come to donate samples, they have to get retested regularly, they have to come for an interview once or twice every 60 days—so it's a lot of time. We try to make it worthwhile for them."

For some donors, the money may make a difference, but most people do it for altruistic reasons, OpenBiome staffers found. "I think the financial aspect is a factor, but people tend to point out that their primary reason is to help save patients' lives," says Langner. Many donors are also excited to help drive science along. "The idea of doing something so good by doing something so simple draws a lot of people to it," Osman says. "If you can do something you do every day anyway and it can save someone's life, it's amazing."

Most people donate for a few cycles and eventually move on. Some stick with it for over a year or until they relocate somewhere. "We have a few donors who have been with us for quite some time," Osman says. "They just keep doing it."

Baldwin planned to be a long-term donor, but at his retesting— the regular 60-day checkup—his blood test returned with slightly elevated levels of two enzymes, AST and ALT, aspartate transaminase and alanine transaminase. These enzymes may spike when a person's muscles suffer some slight damage and breakdown after working out too much. However, highly elevated levels of AST and ALT can be a signal of rhabdomyolysis, a severe and dangerous condition in which dead muscle fibers release their content into the bloodstream, littering it to the point that the kidneys can no longer filter out the toxins. That can cause kidney failure and death.

In Baldwin's case, his enzymes were elevated so slightly that they wouldn't have raised a red flag for his personal physician. But his results were not acceptable under OpenBiome's criteria. While Baldwin had passed the bar the first time, he unfortunately failed his retesting. He couldn't be a donor anymore. Moreover, he couldn't come back and try again a few months later, either.

For Baldwin, his results were an unwelcome surprise. He even talked to a doctor friend to understand why a minor sports injury would disqualify him from being a donor. His friend wasn't sure either. "My doctor friend thought that in theory things like this shouldn't have any impact on your stool," he says. In fact, had he taken this fateful blood test two weeks earlier or two weeks later, his enzyme levels may have appeared as perfect as they were before. But OpenBiome erred on the side of caution. You just can't take any chances, Osman explains. You can't cure C. diff in people and accidently give them some health problem they never had before.

"It's a complex new part of medicine that isn't completely understood, so they are just covering all the bases," Baldwin recalls. The "feel good Friday" emails still remain in his inbox. "It was a cool thing working with them," he says. "If I could do it again, I absolutely would."

Fecal Fortunes and the Future of FMT

When the FDA made the decision to exercise enforcement discretion over fecal transplants, rather than requiring an IND, it indicated that this was a temporary solution. FMT wasn't meant to be the standard of care. It was a stopgap measure put in place until scientists figured out a way to transfer human microbiomes between humans without also transferring their feces. Ideally, patients would receive "clean" bacteria grown in laboratory cultures rather than in someone's feces. That method would make this microbial mix identical to a drug, which could go through the appropriate FDA-required trials to eventually become an FDA-approved medicine.

Several companies found this idea a promising business proposition. So did Smith. Even before launching OpenBiome, he toyed with the idea of making therapeutic bacterial cocktails, aimed to tackle specific diseases. One of the current medical hypotheses was that certain conditions, including autoimmune diseases, are caused by a lack of specific microbial communities in the patient's gut, which either exist in insufficient amounts or are missing entirely. In these cases, transplanting the entire intestinal microbiome may not be necessary. So, unlike FMT, which transfers the donor's entire microbial zoo into the recipient, these cocktails could selectively deliver certain microbial strains to patients who lacked them. The challenge was in mapping microbial imbalances to the diseases they caused. Smith's training as a computational microbiologist was well fitted for the task.

Smith tried a few collaborations with other interested parties, but eventually resolved to form his own company. He had an issue with the approach the other companies were using to do their research. They were going about it the old way, which in his opinion was too slow while not necessarily adding a lot of value.

The traditional drug development paradigm involves either devising a therapeutic molecule or scanning the already available library of compounds to find one that may work for a specific health issue. The next step is to test this new drug in mice for efficacy and side effects—and if all goes well, move on to human trials. This pathway is lengthy and costly, but it's necessary to ensure safety, because many synthetic molecular compounds are novel to the human body and may not be compatible with it.

Accustomed to this drug development trajectory, pharmaceutical companies were following the same pathway when it came to testing stool microorganisms. "Let's take a big library of bacteria, screen it, put it into a mouse, and then eventually put it into humans," Smith explains. "I thought it was a very reductionist approach."

Smith's point was that the many fecal transplants that had already been performed had proved the safety and efficacy of transferring the entire intestinal microbiome in humans. Therefore, isolating individual bacteria or combinations to test them in mice would be setting the research back rather than moving it forward. Mice aren't ideal models for studying human drugs, Smith points out—they are just the best models we have. A lot of drugs fail at the mouse-to-human leap, some because they don't work and others because they produce toxic side effects stemming from the fact that human bodies aren't accustomed to processing these novel, unfamiliar compounds.

But poop isn't like that. "As raw material, stool in principle shouldn't be toxic to us because we carried all this bacteria in us since before we were humans," Smith notes. So as long as the transplants are pathogen-free, there's no risk of toxicity to the liver, kidneys, or other organs. "And in practice, it has been safely administered to over 45,000 patients," he adds. "So by now we have both the theoretical and the empirical safety understanding of this intervention. At this point, mice aren't adding a lot of value. They don't necessarily accurately predict what happens in humans, anyway."

So Smith resolved to form his own company, Finch Therapeutics, with the motto "Humans First." Named after Darwin's finches, known for the remarkable diversity of their beaks that is vital to their survival, the company begins its scientific inquiry in the clinic rather than in the lab. It uses interventional studies (that is, studies done on humans) to understand how microbiome changes affect health, and it uses machine learning to analyze clinical data, zeroing in on particularly beneficial microbial strains. This method also lets Finch determine when a full-spectrum microbiome transplantation is necessary and when a microbial cocktail made with just a few specific strains would be better. In more scientific terms, the company describes this "cocktail" as rationally selected microbiota therapy.

Rationally selected microbiota therapy holds great promise when it comes to diseases originating in specific microbiome imbalances, which medics call dysbiosis. Unlike *C. diff*, caused by a single-pathogen takeover, other gut malfunctions, like ulcerative colitis and Crohn's disease, stem from a lack of certain bacterial strains. For example, people with ulcerative colitis seem to carry low numbers of a bacterial species named *Faecalibacterium prausnitzii*, which produces a necessary intestinal compound, butyrate. People with Crohn's also run low on *F. prausnitzii*, plus they harbor an inflammatory strain of *E. coli*.

Researchers found that just about any healthy donor's stool can restore the bacterial community after a *C. diff* infection. But patients with Crohn's and ulcerative colitis did better with transplants from specific donors whose stool was particularly rich in the specific microbes they lacked. These microbes, dubbed "keystone species," probably produce chemicals whose absence contributes to the disease. The super-donors, colloquially christened "super-poopers," tend to have high levels of these keystone species. Therefore, they may have the ability to cure these specific diseases with their stool. That's one type of therapeutic Finch is working on. "We are using a cocktail of bacteria isolated from super-donors that works really well in ulcerative colitis," Smith says. "And we are now developing a method to treat IBD and Crohn's disease." Named FIN-524, the company's microbial cocktail for ulcerative colitis is the first among Finch's product candidates based on the company's "humans first" philosophy.

But Finch's fast-forward strategy doesn't sit well with other players in the field who are also trying to cook up such bacterial mixes, but via more traditional approaches. These therapeutics have the potential to make huge amounts of money, so the competition is heating up. Some people in the field have christened this new frontline "the poop wars."

Recently, three companies, Rebiotix, Inc., Seres Therapeutics, and Vedanta Biosciences, all of which are forging their own versions of microbial cocktails, formed a coalition called the Microbiome Therapeutics Innovation Group. Their joint website explains that the coalition works to "enhance the regulatory, investment, and commercial environment to accelerate microbiome therapeutic drug product development and expand availability of life-changing and life-saving FDA approved microbiome therapies to patients." But patients and medics working in this field think the group wants to tighten the regulations the FDA loosened in 2013, and to rein in the competition—Finch Therapeutics and OpenBiome.

Because the two sister companies soared to success thanks to the FDA's relaxed regulations, the newly formed coalition sees them as unfair competition. The trio, dubbed by some the "poop drug cartel," wants the regulations back, which could impede Finch's accelerated approach and stymie OpenBiome's ability to supply patients with affordable, life-saving medicine.

Like most pharmaceutical or biotech companies, the trio had to attract investors to conduct their testing and research. So did Finch. At some point, all companies' backers want to see a return on their investments. But the trio followed the traditional route of lengthy and costly trials, gathering data and filing paperwork to document outcomes and side effects. And it's hard to compete with a rival that plays by a different, easier set of rules. "These companies found it difficult to conduct traditional clinical trials because who wants to bother with all this research paperwork when there's a product already available?" Khoruts says, referring to OpenBiome's service. For as long as the FDA allows the enforcement discretion loophole, following the traditional path puts the trio at a competitive disadvantage.

And so, after a "lot of back and forth at meetings and expressions of annoyance," the trio formed their coalition and moved into action. "Their site says explicitly to 'engage the policymakers,'" Khoruts points out, "which people read as 'companies got together to apply pressure on the FDA to shut down the OpenBiome operation.'"

That prompted an outcry from patients scared that the straightforward and affordable *C. diff* cure might soon disappear. At this writing, OpenBiome charges $1,595 for its lower-delivery preparation and $1,950 for its capsules. Certain components of the FMT process are covered by some health insurance plans. Moreover, for people who can't afford FMT, OpenBiome covers the cost of the material.

Overall, even without insurance, the FMT price isn't going to break the bank—compared with the cost of vancomycin, it is a minuscule amount. But we are all too familiar with the astronomical prices of new drugs that enter the market when there's little competition.

"These companies are racing in Phase III clinical trails and getting the FDA approval, and there's a danger that they can get exclusivity on the market," Khoruts says, describing the fears of patients who are following this policy war. They are worried that business interests will supersede patients' needs. "You can be sure that there will be another drug that would cost thousands of dollars," Khoruts adds, "so it will become inaccessible to many patients—either because they don't have insurance and they can't afford it, or they have insurance but it doesn't cover it until you exhaust all cheaper options."

Khoruts argues that even after the new stool-based drugs are released, stool banks should be allowed to exist to keep the process affordable. "Why can't this be like a blood banking model?" he questions. "Transfusions are a more stringent paradigm compared to drugs," he points out, but regulations on FMT can be put in place to preserve patients' interests. "Why are we giving in to these companies?"

This fecal saga is to be continued. The coalition will try to push for more regulations. But patients and providers intend to fight back. Last but not least, Carolyn Edelstein and Mark Smith formed their own coalition—a new and powerful one. They got married. And they are certainly not retreating from the fecal battlefield.

More than 300 years later, we are back to the poop wars. Only prices no longer amount to three *bu* of silver or half a *ryo* of gold, as in eighteenth-century Osaka, but rather millions of millions of dollars. And the wars aren't waged within geographic locations, but via capsules and bottles, with contestants competing for the chance to cure diseases and vanquish superbugs. Conceptually, however, we are nearing the thinking of our Chinese and Japanese ancestors, down to the understanding that all feces are not created equal, but differ in their nutritional and healing power. Like the medieval *fenfu* men who treasured the city areas with healthier populations producing richer night soil, medics now value our twenty-first-century superdonors. And now not only scientists, but painters and multimedia and performance artists, are joining to celebrate that fact.

AFTERWORD: BREATHING POETRY INTO POOP

"I have a confession to make," New York performance artist Shawn Shafner declares from the stage. "I can't keep it inside anymore." Holding his hands to his chest, he rips open his white dress shirt in a dramatic gesture—and reveals a T-shirt with big letters splattered on the front. Just in case someone in the audience can't see the statement, he screams out his eerie affirmation: "I am a POOPer!"

The audience cracks up. Shafner builds on the momentum with a few more bathroom jokes, and the laughter grows into a roar. The creator of The POOP Project—the acronym stands for People's Own Organic Power—Shafner is infinitely creative with his toilet puns, but he is also serious about the underlying science. In his performance, he explains the earth's nitrogen cycle, the problems of modern sewers, and the inner workings of our guts.

Shafner's relationship with his own organic power is a complicated one. As a child, he was ultimately disgusted by it. He found it so appalling that he would sit on the toilet and squirm to tighten his sphincter. "I would sit there and tell my poop, 'no, you are not coming out today.'"

He eventually outgrew his disgust, and that's why he started his performances. He uses his unconventional shows to start a conversation about a topic that still makes many people wince. Even this book struggled with the yuck factor. Several publishing houses turned it down because they felt uncomfortable with the topic—even though they found it interesting. Editors said they would feel awkward discussing the subject at editorial meetings. They didn't know how to present the idea to their marketing departments. They couldn't imagine what they would put on the cover.

People talk eagerly about how they slice veggies, roast meat, mix sauces, and whip cream for a party, yet hardly anyone discusses what happens to all that goodness on the other end. That is a societal problem. But our unwillingness to talk about it creates a cascade of other problems—medical, biological, environmental. In order to deal with all these issues, we first have to learn to talk about poop.

That's what Shafner wants his audience to do. "We are all POOPers," he tells them. The sooner we admit it, the better we will handle this complicated substance. And because nothing breaks taboos better than humor, he goes for the hilarious grand finale. "Sing with me," he calls out to the audience with wide, outstretched arms. "I am a POOPer, I am a POOPer!"

Massachusetts-based artists Caitlin Foley and Misha Rabinovich take a different position on human dark matter. They want to breathe poetry into poop.

Combining drawing and sculpture with multimedia techniques, Foley and Rabinovich create art installations that teach people about gut health, intestinal bacteria, and the microbiome. The duo weaves these scientific concepts into the cultural landscape of society with crafty metaphors, analogies, and allegories.

"We use bacterial culture as the analogy of human culture because human culture itself is a little bit like the microbiome." Foley says. "It needs diversity." Isolated, reclusive monocultures become intolerant and unsustainable. And just as a rich, diverse microbiome makes a stronger human, a rich, diverse culture makes a resilient and prosperous society.

This point was demonstrated when the artists helped OpenBiome create a map of the company's stool sample shipments within the United States. After they built the map, they found that the blue states requested a lot more shipments than the red ones. The blue states are also more densely populated, so it was hard to pinpoint the exact causes of this difference, but it could be that people in the liberal enclaves were more open to trying a new, unconventional treatment than those in conservative locales. That might leave patients outside the blue areas at a disadvantage when it comes to battling superbugs.

In order for people to fully benefit from our dark matter—whether as medicine, fertilizer, or renewable energy source—it must be cleared of its stigma. And that's what Foley and Rabinovich aim to do. They want to take human waste's new identity out of the research labs and to the masses. "Waste is a human construct," Rabinovich says. It's something we produce but have no use for. But as this book has shown, people's own organic power has numerous uses—and therefore our view of it can, and should be, changed.

Art can help destigmatize it. Science is constrained by the necessity of describing substances and processes with great precision, but art can help people make the leaps of imagination necessary to push entrenched cultural, societal, or hygienic limits—especially when those limits no longer make sense. That's where humor comes in handy once again. Foley and Rabinovich hope that the smiley, cape-flaring Super Turd character they created for their installations can help people make that leap.

Humans' attitudes toward their dark matter are changing. In 2018, a Vienna-based art collective named Gelatin created an exhibit in which people walked through excrement—not real, but a set of four gigantic specimens of people's own organic power. The artists' creations didn't exactly correspond to the Bristol Stool Chart's specimens, with their "clear-cut edges" or "cracks on the surface," but they were fully realistic. The huge brown-black, anatomically correct glops looked like they had just emerged from an enormous bowel. More interestingly, visitors were encouraged to walk through the exhibit in nude suits available at the coat check. Why nude suits? It's simple, the artists say. You can often tell by the clothes who a person is—a banker, for example. But nakedness makes everyone equal. And everyone poops. Moving bowels is one big equalizing commonality that unites humankind regardless of ethnicity, color, religion, diet, or traditions. When I visited the Sulabh International Museum of Toilets in New Delhi, India, which chronicles the development of these facilities throughout history, its curator, Manoj Kumar, talked about how its exhibits touch every visitor, no matter the country or conti-

FIGURE 8. Manoj Kumar, curator at the Sulabh International Museum of Toilets in New Delhi, India, holds a piece of decomposed human waste in his hand, demonstrating how the Sulabh flush toilet converts sewage to safe fertilizer. CREDIT: LINA ZELDOVICH

nent they come from. "It is one thing we all have in common," he said, standing next to a prototype of a composting toilet that the Indian government offers to subsidize for farming families.

More refreshingly, poop is making its way into pop culture, too. Poop emojis are everywhere these days. A cute poo pillow was a best seller on Amazon a couple of years ago. During the 2018 holiday season I bought a pom-pom hat with a sequined poo pile in a children's clothing store. The sequins were reversible, so I could flaunt my poo in two different ways—shimmering silver or radiant rainbow. My friend's daughter was shocked to see me proudly sporting that hat, but five minutes later she wanted one, too. The hat wasn't the only accessory of the kind. The store was brimful of colorful poo splotches sparkling on girls' shirts.

Perhaps this generation won't be ashamed of their organic power. They won't think of it as waste. Instead, they will be able to discuss its applications in medicine, agriculture, and energy with ease and care.

In fact, my son proudly wore the "I am a POOPer" T-shirt to school on World Toilet Day. When he explained the shirt's meaning in his science class, his friends deemed it a cool statement. Perhaps as they grow up and start watching the Super Bowl together, they will also deem it cool to celebrate the National Poop Day the Monday after.

I am glad that humans are reawakening to their organic power. I am glad that artists and scientists have already warmed up to it. I don't know which way humankind will go in its poop upcycling—the Lystek, Cambi, biochar, biogas, or biocrude method—but I am glad that we now have solutions that can work in many different settings, from very rural to densely urban and from personal to industrial. What I do know, though, is that in an era when pharmaceutical companies are arguing over poop-related patents, it's imperative that the general public joins in the debate. It's important that we do so because poop is part of our future health care. It is also part of our future environment. But for these debates to occur, the public, policymakers, and businesspeople must be able to discuss the subject easily and genuinely. It may take us some time to reach this dark matter nirvana, but perhaps poop puns, poop poetry, and poop pom-pom hats can lead the way.

I love my poop hat. My grandfather would have loved it, too. And he would have sung along with Shafner. He would have walked through the excrement exhibit with me, laughing like a child. I could just hear him cracking a joke, "So my butt creates art, eh?" And the next fall, he would have donned his gray overalls and proceeded to empty the septic system onto our land once again. He didn't get to see humankind catch up with his wisdom, but I am starting to witness it.

So let's just all of us say, "We're POOPers," and let go of the stigma. Let's give our excrement a little more breathing room. Let's make our poop a little more poetic. Let's pat ourselves on the back for the amazing potential we carry within. And most importantly, let's not just sit there and watch this versatile, renewable power go to waste. Especially since we now know its worth.

NOTES

Chapter 1

1. "Top Toys for Christmas 2020—List of Best Toys," ToyBuzz, accessed December 10, 2020, https://toybuzz.org/top-toys-for-christmas/.

Chapter 2

1. Charles Ginenthal, *Pillars of the Past*, vol. 4, *Chronology of the Age of Stonehenge and the Megalithic World* (Ivy Books, 2012), 370.
2. Steven Mithen, *Thirst: Water and Power in the Ancient World* (Cambridge, MA: Harvard University Press, 2012), 77.
3. W. J. Corrigan, "Sanitation under the Ancient Minoan Civilization," *Canadian Medical Association Journal* 27, no. 1 (1932), 77–78.
4. A. N. Angelakis, D. Koutsoyiannis, and G. Tchobanoglous, "Urban Wastewater and Stormwater Technologies in Ancient Greece," *Water Research* 39, no. 1 (January 2005), 210–220, https://doi.org/10.1016/j.watres.2004.08.033.
5. Harold Farnsworth Gray, "Sewerage in Ancient and Medieval Times," *Sewage Works Journal* 12, no. 5 (September 1940), 942.
6. Gray, "Sewerage in Ancient and Medieval Times," 939–946.
7. M. Jansen, "Water Supply and Sewage Disposal at Mohenjo-Daro," in "The Archaeology of Public Health," special issue, *World Archaeology* 21, no. 2 (October 1989), 177–192, https://doi.org/10.1080/00438243.1989.9980100.
8. Ann Olga Koloski-Ostrow, *Archaeology of Sanitation in Roman Italy: Toilets, Sewers, and Water Systems* (Chapel Hill: University of North Carolina Press, 2015), 86.
9. Koloski-Ostrow, *Archaeology of Sanitation in Roman Italy*, 1–101.
10. Koloski-Ostrow, 89.
11. Koloski-Ostrow, 1–101.
12. Koloski-Ostrow, 1–101.
13. Koloski-Ostrow, 63.

Chapter 3

1. Kayo Tajima, "The Marketing of Urban Human Waste in the Early Modern Edo/Tokyo Metropolitan Area," *Urban Environment* 1 (2007), 3.

2. Susan B. Hanley, "Urban Sanitation in Preindustrial Japan," *Journal of Interdisciplinary History* 18, no. 1 (Summer 1987), 1–26.

3. Hanley, "Urban Sanitation in Preindustrial Japan," 1–26.

4. Philipp Franz von Siebold, *Manners and Customs of the Japanese, in the Nineteenth Century: From the Accounts of Dutch Residents in Japan and from the German Work of Philipp Franz von Siebold* (North Clarendon, VT: Tuttle, 1981), 329.

5. Tajima, "Marketing of Urban Human Waste," 9–10.

6. Tajima, 2.

7. Hanley, "Urban Sanitation in Preindustrial Japan," 1–26.

8. Tajima, "Marketing of Urban Human Waste," 8.

9. Hanley, "Urban Sanitation in Preindustrial Japan," 1–26.

10. Hanley, 1–26.

11. Alan Macfarlane, "The Excremental Chain" (copyright 2002), 1–29, http://www.alanmacfarlane.com/savage/DUNG.PDF.

12. Hanley, 1–26.

13. Tajima, "Marketing of Urban Human Waste," 2.

14. Macfarlane, "Excremental Chain," 1–29.

15. Ronald P. Dore, *Shinohata: A Portrait of a Japanese Village* (Berkeley: University of California Press, 1994).

16. Hanley, "Urban Sanitation in Preindustrial Japan," 24.

17. Donald Worster, "The Good Muck: Toward an Excremental History of China," *RCC Perspectives: Transformations in Environment and Society*, no. 5 (2017), https://doi.org/10.5282/rcc/8135.

18. Rose George, *The Big Necessity: The Unmentionable World of Human Waste and Why It Matters* (New York: Metropolitan Books, 2008).

19. Gene Logsdon, *Holy Shit: Managing Manure to Save Mankind* (White River Junction, VT: Chelsea Green, 2010), 5.

20. Worster, "The Good Muck."

21. Logsdon, *Holy Shit*, 5.

22. Justus von Liebig, *Chemistry in Its Application to Agriculture and Physiology*, 4th American ed. (Cambridge, MA: John Owen, 1843), https://archive.org/stream/chemistryinitsap00lieb/chemistryinitsap00lieb_djvu.txt.

23. Richard Jones, "Manure and the Medieval Social Order," in *Land and People: Papers in Memory of John G. Evans*, vol. 2, ed. Michael J. Allen, Niall Sharples, and Terry O'Connor (Oxford: Oxbow Books, 2009), 215–224.

24. Harold Farnsworth Gray, "Sewerage in Ancient and Medieval Times," *Sewage Works Journal* 12, no. 5 (September 1940), 939–946.

25. Gray, "Sewerage in Ancient and Medieval Times," 939–946.

26. Gray, 939–946.

27. Laurence Wright, *Clean and Decent: The Fascinating History of the Bathroom and the Water-Closet and of sundry habits, fashions & accessories of the toilet principally in Great Britain, France, & America* (London: Routledge & Kegan Paul, 1960), 145.

28. *The Outlines of Flemish Husbandry: As Applicable to the Improvement of Agriculture in Canada; Originally Published by the Society for the Diffusion of Useful Knowledge, and Re-Published by the Bureau of Agriculture, in French and English* (London: Forgotten Books, 2018), 1–146.

29. Macfarlane, "Excremental Chain,"1–29.

30. Amy Bogaard, "Middening and Manuring in Neolithic Europe: Issues of Plausibility, Intensity and Archaeological Method," in *Manure Matters: Historical, Archaeological and Ethnographic Perspectives*, ed. Richard Jones (London: Routledge, 2012), 25–40.

31. Ellen Castelow, "The Throne of Sir John Harrington," Historic UK, accessed December 11, 2020, https://www.historic-uk.com/CultureUK/The-Throne-of-Sir-John-Harington/.

32. D. H. Craig, *Sir John Harington* (Boston: Twayne, 1985); Jason Scott-Warren, "The Privy Politics of Sir John Harington's "New Discourse of a Stale Subject, Called the Metamorphosis of Ajax," *Studies in Philology* 93, no. 4 (Autumn 1996), 412–442.

33. Karen Bernier-Cast, "Gargoyles, Kisses and Clowns: A Study of Carnivalesque Male Urinals and Restrooms," *Material Culture* 43, no. 1 (Spring 2011), 21–39.

34. Tony Dingle, "The Life and Times of the Chadwickian Solution," in *Troubled Waters: Confronting the Water Crisis in Australia's Cities*, ed. Patrick Troy (Canberra: ANU Press, 2008), 10.

Chapter 4

1. Edwin Chadwick, *Commentaries on the Report of the Royal Commission on Metropolitan Sewage Discharge, and on the Combined and the Separate Systems of Town Drainage* (London: Longmans, Green & Co., 1885), 1–42.

2. Richard Adler and Elise Mara, *Typhoid Fever: A History* (Jefferson, NC: McFarland, 2016), 67.

3. William Budd, "Typhoid Fever in its Nature, Mode of Spreading and Prevention," chapter 14 in *Public Health: The Development of a Discipline*, vol. 1, *From the Age of Hippocrates to the Progressive Era*, ed. Dona Schneider and David E. Lilienfeld (New Brunswick, NJ: Rutgers University Press, 2008), 482, https://books.google.com/books?id=4xv0scbWSAUC&pg=PR5&dq=%E2%80%9CIndia+is+in+revolt+and+the+Thames+stinks!&source=gbs_selected_pages&cad=3#v=onepage&q=%E2%80%9CIndia%20is%20in%20revolt%20and%20the%20Thames%20stinks!&f=false.

4. Chadwick, *Commentaries on the Report of the Royal Commission*, 1–42.

5. David Edward Owen, *The Government of Victorian London, 1855–1889* (Cam-

bridge, MA: Belknap Press of Harvard University Press, 1982); Chadwick, *Commentaries on the Report of the Royal Commission*, 1–42.

6. Chadwick, 1–42; Christopher Hamlin, "Edwin Chadwick and the Engineers, 1842–1854: Systems and Antisystems in the Pipe-and-Brick Sewers War," *Technology and Culture* 33, no. 4 (October 1992), 680–709, https://doi.org/10.2307/3106586.

7. Chadwick, *Commentaries on the Report of the Royal Commission*, 1–42.

8. Chadwick, 1–42.

9. Justus von Liebig, *Chemistry in Its Application to Agriculture and Physiology*, 4th American ed. (Cambridge, MA: John Owen,1843), 194, https://archive.org/stream/chemistryinitsap00lieb/chemistryinitsap00lieb_djvu.txt.

10. Joel A. Tarr, "From City to Farm: Urban Wastes and the American Farmer," *Agricultural History* 49, no. 4 (October 1975), 598–612, https://www.jstor.org/stable/3741486.

11. Tarr, "From City to Farm," 598–612.

12. Tarr, 598–612.

13. Tarr, 598–612.

14. Chadwick, *Commentaries on the Report of the Royal Commission*, 12.

15. Chadwick, 1–42.

16. Chadwick, 1–42.

17. Chadwick, 1–42.

Chapter 5

1. Howard Gest, "The Discovery of Microorganisms by Robert Hooke and Antoni van Leeuwenhoek," *Notes and Records of the Royal Society of London* 58, no. 2 (May 2004), 187–201, https://doi.org/10.1098/rsnr.2004.0055.

2. Gest, "Discovery of Microorganisms, 187–201.

3. Gest, 187–201.

4. S. Ramaseshan, "The Amoral Scientist—Notes on the Life of Fritz Haber," *Current Science* 77, no. 8 (1999), 1110–1112, http://www.jstor.org/stable/24103592.

5. Ramaseshan, "The Amoral Scientist," 1110–1112; Keith L. Manchester, "Man of Destiny: The Life and Work of Fritz Haber," *Endeavour* 26, no. 2 (June 2002), 64–68, https://doi.org/10.1016/s0160-9327(02)01420-5.

6. Vaclav Smil, *Enriching the Earth: Fritz Haber, Carl Bosch, and the Transformation of World Food Production* (Cambridge, MA: MIT Press, 2004).

7. Donald Worster, "The Good Muck: Toward an Excremental History of China," *RCC Perspectives: Transformations in Environment and Society*, no. 5 (2017), https://doi.org/10.5282/rcc/8135.

8. Daniel Max Gerling, "American Wasteland: A Social and Cultural History of Excrement, 1860–1920" (PhD thesis, University of Texas at Austin, 2012), 80, http://hdl.handle.net/2152/ETD-UT-2012-05-5036.

9. Gerling, "American Wasteland," 269–274.

10. Gerling, 275–276.
11. Centers for Disease Control and Prevention, "Global Diarrhea Burden," Global Water, Sanitation, and Hygiene (WASH), last reviewed December 17, 2015, https://www.cdc.gov/healthywater/global/diarrhea-burden.html.
12. Rose George, *The Big Necessity: The Unmentionable World of Human Waste and Why It Matters* (New York: Metropolitan Books, 2008).
13. Bindeshwar Pathak, *New Princesses of Alwar: Shame to Pride* (New Delhi: Sulabh International Social Service Organisation, 2009).

Chapter 9

1. https://www.who.int/news-room/fact-sheets/detail/household-air-pollution-and-health.

Chapter 12

1. Edwin Chadwick, *Commentaries on the Report of the Royal Commission on Metropolitan Sewage Discharge, and on the Combined and the Separate Systems of Town Drainage* (London: Longmans, Green & Co., 1885), 1–42.

Chapter 13

1. "Factbox: How Much Is 60 Million Barrels of Oil?" *Reuters Business News*, June 23, 2011, https://www.reuters.com/article/us-iea-oil/factbox-how-much-is-60-million-barrels-of-oil-idUSTRE75M6S520110623.
2. "How Much Oil Is Consumed in the United States?" FAQ, US Energy Information Administration, https://www.eia.gov/tools/faqs/faq.php?id=33&t=6.

Chapter 14

1. Melinda Wenner Moyer, "Gut Bacteria May Play a Role in Autism," *Scientific American Mind* 25, no. 5 (September 1, 2014), https://www.scientificamerican.com/article/gut-bacteria-may-play-a-role-in-autism/.
2. Qinrui Li, Ying Han, Angel Belle C. Dy, and Randi J. Hagerman, "The Gut Microbiota and Autism Spectrum Disorders," *Frontiers in Cellular Neuroscience* 11 (2017), https://doi.org/10.3389/fncel.2017.00120.
3. Li, Han, Dy, and Hagerman, "Gut Microbiota and Autism Spectrum Disorders."
4. Alexander Khoruts, "Fecal Microbiota Transplantation: An Interview with Alexander Khoruts," interview by Steve Lebeau, *Global Advances in Health and Medicine* 3, no. 3 (May 1, 2014), 73–80, https://doi.org/10.7453/gahmj.2014.020.
5. Faming Zhang, Wensheng Luo, Yan Shi, Zhining Fan, and Guozhong Ji,

"Should We Standardize the 1,700-Year-Old Fecal Microbiota Transplantation?" *American Journal of Gastroenterology* 107, no. 11 (November 2012), 1755, https://doi.org/10.1038/ajg.2012.251.

Chapter 15

1. Lawrence A. David, Arne C. Materna, Jonathan Friedman, Maria I. Campos-Baptista, Matthew C Blackburn, Allison Perrotta, Susan E. Erdman, and Eric J. Alm, "Host Lifestyle Affects Human Microbiota on Daily Timescales," *Genome Biology* 15, no. 7 (2014), 2–15, https://doi.org/10.1186/gb-2014-15-7-r89.
2. Elie Metchnikoff, *The Prolongation of Life: Optimistic Studies* (New York: G. P. Putnam's Sons, 1910), 96.

INDEX

Italic page numbers refer to photographs.